First published in 20.. by PRESS DIONYSUS LTD in the
UK, ... Copthall Road, W1 5BZ, London.

www.pressdionysus.com

Paperback

ISBN 978-1-913961-46-6
Copyright © 2035 by PRESS DIONYSUS LTD.

DIONYSUS

First published in 2025 by PRESS DIONYSUS LTD in the UK, 167, Portland Road, N15 4SZ, London.

www.pressdionysus.com

Paperback

ISBN: 978-1-913961-46-6

TO BE READY FOR THE FUTURE WITH AI

Mustafa Efe TOPALOGLU

DIONYSUS

ISBN- 978-1-913961-46-6

© Press Dionysus 2025

English Translation by Mesut ŞENOL

Press Dionysus LTD, 167, Portland Road, N15 4SZ, London
- e-mail: info@pressdionysus.com
- web: www.pressdionysus.com

ABOUT THE AUTHOR

Mustafa Efe Topaloğlu was born as a child of a household with 13-person family members in Çayeli district of Rize Province. Their great grandfathers had migrated to Rize from Hopa/Artvin, and Topaloğlu started his education in the elementary school of Aşıklar Village of Çayeli district. Following his graduation from Çayeli Religious Vocational Junior High School and Çayeli High School, he earned his diploma from the Business Administration Faculty of the Anatolian University.

When only just 13 years old, he got to Istanbul and started working at an advertisement agency. At 15 he returned to his hometown to set up his own advertisement agency, and commenced his business life. Topaloğlu resumed on his education. In the following years, again in a very young age he made use of his experience he accumulated during his agency work to put up the first local radio of the Black Sea Region called Çay FM, and went on television broadcasting business through new initiatives in the media sector. Moving to Istanbul in 1994, Mustafa Efe Topaloğlu, took an active role in the media sector there by owning and managing a number of sectoral newspapers and magazines. He also was at play in the fields of music production and event organization.

Setting foot in the software sector in 2004, Topaloğlu accomplished projects in the fields of e-commerce, software and cinematics. He is the producer of a feature film called "I am Nothing". He developed original application in the software sector. He owns a couple of useful model patents. Topaloğlu has been assuming active duties in NGOs, performing the roles of presidency of these organizations or sitting on their boards. Currently he serves on the Advisory Board of the Economics, Administrative and Social Sciences Faculty of Istanbul Gelişim University.

An entrepreneur businessman Mustafa Efe Topaloğlu currently runs his companies operational in the fields of software, film, agency services, and internet media sectors as being their CEO.

CONTENTS

VIII. TRANSPORTATION, TRAVEL AND SAFETY IN THE FUTURE

IX. MEDIA, COMMUNICATIONS AND CULTURAL CHANGES

X. FUTURE ENVIRONMENTAL POLICIES AND RELATIONSHIP WITH NATURE

XI. AI AND GLOBAL POLICIES

INTRODUCTION

While writing this book, I studied hundreds of research, and maybe dozens of articles. I was planning to share the scientific data I reached at the end of three-year-long work I was working on. Nevertheless, within the process I realized that AI would reveal successful and dependable results in research theme.

I noticed that the analyses presented by AI, to which I from time to time asked questions, happen to be as much scientific and credible as the articles I read and the investigations I made. That's why I made my mind that I should write this book solely with the help of AI.

Another reason I wrote this book with the assistance of AI is to raise an awareness in this subject, and to demonstrate what AI could accomplish, its legal dimensions, and how much it would be part of our life in the future. In today's world, millions and even billions of people are wondering how the future of humanity would be shaped. Even though the scientific studies and research about the future are quite deep, yet there are not as many scientific study and research tools or organizations to analyze the future with a broad and objective perspective as much as AI could.

Which questions are being asked today by people to AI put into use in many parts of the world? How will AI affect our lives in the future? What kind of transformation is the world going through? Which professions will be less important, and which ones will become the professions of the future? Which professions will appear? Which ones will earn the most? And which ones will present more career opportunities? The unemployed, casualties of AI, would be doing what? In which fields the people without a diploma would be successful in the future? Such and such questions make up the backbone of this book. I have written this book entirely with the help of AI in order to find answers to these questions and demonstrate the boundaries of AI. This book is not just a research study but at the same time it has the quality of a guide to address the issues related to the future through the lens of AI. I would like to extend my gratitude in advance for your accompanying me in this journey.

Throughout history of humanity, knowledge, skills and capabilities have been passed down through generations, and every era in history had been shaped according to their unique technological revolutions. Works related to the abilities of humans to be transferred to machinery were conducted from 1950s on, gathered speed thanks to the development of the information technologies in recent years, and reached a level of creating transformation in many aspects of life.

While nowadays AI technologies keep on contributing to the growth of the economies of the countries on a macro scale, and they also became a strategical element increasing the productivity of the businesses on the macro scale. On the one hand, the states utilize the investments they make with the help of these technologies as a critical move for the future, the business world enhances new adaptation patterns to seize the opportunities presented by AI, on the other.

In the near future, AI will go beyond assuming a role in facilitating human life, and it will generate revolutionary changes in many sectors such as education, health, transportation, agriculture, security, law, economics, food, and energy. Smart cities, humanoid robots, digital human bodies and artificial organs produced thanks to biotechnology are among the breakthroughs to have a huge impact on human life. In this process, in order to keep up with the global competition, it is vitally important to be able to adapt to AI technologies.

However, as much as the opportunities presented by AI, we also face the critical issues of reigning over AI. The legal and ethical dimensions of AI technologies require the integration of some regulations with the internal control mechanisms of the organizations. Otherwise, the use of out-of-control AI could do more harm than good.

As this book is examining the effects of AI on the business world and human life, at the same time, from an ethical and legal framework it offers some foresight into how the relationship between AI and humans will be shaped. It aims to make a contribution to the people who are going to direct these technologies in the future world, in understanding how to manage this transformation.

I wish that AI technology could serve best to humanity in the future…

DEVELOPMENT AND HISTORY OF AI

As the Artificial Intelligence – AI happens to be a bourgeoning field since 1950s, especially in the last 10-15 years, it made great strides. We sort its important milestones by the following short headings:

The Birth of AI Concept in 1950s: By posing a question of "Could machines think?", Alan Turing developed the concept of so-called "Turing Test". The term of AI was officially used for the first time during Dartmougth Conference in 1956.

The AI Winter Between 1970 and 1990: As the anticipated developments were not actualized, the interest of investors decreased. Only in 1980s, the specialized systems became popular and some companies invested in this field.

The Rise of Big Data and Deep Learning in 2000s: With the computers gaining greater capability, Big Data started to feed AI. Giants such as Google, IBM, and Microsoft began to make serious investment in AI.

The Era of Deep Learning and AI in 2010s: The success earned by Alex-Net in the image recognition contest in 2012 paved the way for deep learning. In 2016 Google's software AlphaGo, beat the world Go Champion, and this caused a huge splash. OpenAI and Google enhanced the transformer models which laid the foundation of the modern AI systems in 2017.

OpenAI, put ChatGPT in service in November 2022, and altered the interactions of millions of people with AI.

Finally, in 2023, GPT-4 was introduced, and as the more advanced language model, he gained the humanlike text production capability.

AI at the present time

AI revolutionizes the fields of health, finance, education, industry, and even literature. While I was writing this book, I asked AI a question which was never asked before. If AI had its own consciousness, in which fields it would have liked to help humans? The answer AI gave impressed me a lot. "Had I had my consciousness, I would have preferred to work on the project which would increase the happiness, well-being and creativity of humans. To spread the knowledge on an equal basis, to revolutionize health, to provide equal opportunities in education and challenges such as climate change and global issues would be my priorities" it replied. We all talk about AI's dangers and its shortcomings, but in fact, the most important issue here is, to think about how we can use it to the utmost benefit of humanity.

The future of AI will be shaped by the questions we ask and how we will be using it to attain which objectives. The book of *"To Be Prepared for The Future with AI"* is focusing exactly on these questions. This book is written by AI. But it didn't come into being on its own, and AI didn't say "I will write a book". As a prompt engineer, it was me to orient it and ask it right questions and seek for right answers. Just like a master's shaping the material at hand, and create a piece of art out of it. When writing this book, AI and I worked with team spirit. This book was materialized thanks to the questions we wondered about the future.

When we take a look at the history of humanity, we can readily realize that the development always starts with asking questions. Had humans would have not wondered about new things, and asked the right questions, then we would have not reached the point we are now at. AI works according to the same logic. The more you ask it the right and deep questions, the better replies it provides for you. If you pose nonsense questions to it, then the answers it would give you would not be as satisfying as you'd expect.

I don't consider this book merely a book written for the sake of authoring a book. This book means a starting point for me, to understand the future and be prepared for it.

I hope it does give the same motivation to the reader...

I

NEXT CENTURY AI

What Does Being A Human Mean?

This transformation is going to bring along the deepest ethical, philosophical and social questions of the human history: "What does being a human mean?" And, "Where does the sense of self begins and ends?"

Physiological and bodily transformation, biotechnology, nanotechnology, genetic engineering, can materialize through cybernetics (integration of machine and human) in the future. The following are of some likely aspects of this transformation:

Reproduction of biological organs (organ bioprinters)

• Thanks to bioprinters, personalized, artificial and completely functional organs will be produced.

• Organs such as heart, kidney, liver could be quickly transplanted without referring to waiting lists.

Genetic manipulation and CRISPR technolojy

• Through gene regulation techniques, congenital diseases, aging and genetic disorders will be eliminated.

• By rewriting human DNA, stronger and long-lived individuals could be created.

Repair on a cellular level through nanotechnology

• Nano robots to be embedded in the body will repair cells, fight infections and get rid of toxins.

• These robots could slow the aging or stop it completely.

Cybernetic limbs and organs

• Bionic arms, legs, and sensory organs (eye, ear) will give super abilities to humans.

• These limbs will be controlled via thought, even they will be stronger than the physical organs.

Expanding sensory abilities

• Thanks to eye implants, the night vision, infrared detection, augmented reality experiences will be possible.

• Thanks to ear implants, ultra-sensitive sound detection abilities could be improved.

Biological syntesis: Human and animal skills

• Integration of animal genes or biological features of animals to be integrated with humans, like having a sharp vision like an eagle or the skill of a whale being able to hold the breath for a long time could be at the top of the agenda.

Smart skin and tissues

• Smart skin coating which can repair the wound by itself, and sensitive to heat and pressure will be produced.

• "Smart layers" which can be worn on the body or implanted in the body will be monitoring state of health.

Digitalization of the body

• Instead of physical body, digital avatars or "metaphysical bodies" represented by holographic projections could come up.

These developments could not only transform the human body in terms of health, but also radically, as far as aesthetic and performance are concerned. Yet this transform will require reconsideration of the discussions on the "natural" human body concept and ethics.

Emotional and Conscious Algorithms

Emotional and conscious algorithms mean that AI will not only make logical decision but also be able to proceed within emotion and consciousness-like proceses. These kinds of algo-rithms would aim to develop a capacity resembling human-like intuition, empathy and self-awarness.

Emotional algorithm are the systems being able to give ap-propriate reactions by analyzing the emotional states of humans. These algorithms:

• Will trry to recognize the emotions of humans through analyzing facial expressions, tone of voice, and biometric data.

• Will create emphatic reactions and connaturalize human interactions.

Model applications

• **Intelligent assistants:** By noticing the fact that users are stressed or sad, they can employ appropriate way of speaking.

• **Costumer Services:** By recognizing the emotional state, they can exhibit politer and calm approaches.

• **Health technologies:** By recognizing the symptoms of depression and anxiety, they can make supporting recommen-dations.

Conscious algorithms (*Artificial Consciousness*)

Conscious algorithms are the systems aiming at giving the machines sense of self-awareness. The fundamental features of these algorithms are as follows:

• **Self-awareness:** Being aware of its own state.

• **Thought chain:** When making decisions, being able to predict depending on the past experiences.

• **Self-evaluation:** To optimize its own behavior by recognizing its own mistakes.

Suggested applications

• **Autonomous robots:** They can analyze their own duties and develop new strategies.

• **Creative systems:** They can create human-like works in the fields of arts, music or literature.

- **Ethical Decision Mechanisms:** Systems which are able to make ethical decisions in difficult circumstances could be developed.

Challenges and risks

- **Concept of consciosness:** Since the definition of consciousness is not fully comprehended, these kinds of algorithms would induce philosophical and scientific discussions.

- **Ethical issues:** Should the conscious systems be given rights? Can an artificial entity feel pain?

- **Security threats:** Could AI which is able to set its own targets pose danger?

- **Risk of manipulation:** Emotional algorithms could be used to orientate the decision making processos of humans.

Emotional and conscious algorithms have the potential to influence the human-machine interactions deeply. Nevertheless, the etchical, social and security dimensions of these developments should be handled meticulously.

Artificial Biological Intelligence

(*Synthetic Biological Intelligence*)

Artificial biological intelligence indicates the fact that biological systems (cells, neurons, and genetic structures) are artificially designed to perform cognitive functions. This consept delineates the building AI on biological materials by going beyond silicone based sistems.

What makes it work?

Artificial biological intelligence comes out through a merger of biotechnology, neural engineering and genetic engineering.

1. **Biological neural networks:** By integrating digital circuits, live neurals produced in labs could perform data processing functions.

2. **Genetic engineering:** By being genetically reprogrammed, cells could store data or make calculations.

3. Biochips: Chips providing the flow of data between biological structures through electronic circuits could be enhanced.

4. Cellular based calculations: By having some chemical reactions on biological cells, it is possible to go beyond classical algorithms.

Fields of application

1. Advanced calculation

• Artificial biological intelligence could be used to solve complex problems which cannot be figured out by classical computers.

2. Health and treatment systems

• With the brain-chip interfaces, breakthroughs could be achieved in the treatment of neural diseases.

• Analyzing complicated diseases such as cancer on a cellular level through "biological algorithms" could be possible.

3. Autonomous systems

• By being equipped, robots and machines would become fast learners and well adapted in the environment.

4. Data storage and processing

• As the extention of the biological intelligence, DNA based data storage systems enable storing and processing of big amount of data.

Advantages

• **Low energy consumption:** Compared with the classical computers, biological systems consume a lot less energy.

• **High level of adaptability:** Biological systems can quickly adap to environmental conditions.

• **Parallel processing:** Cellular calculations can take many actions simultaneously.

Challenges

• **Controlling issue:** Biological systems can exhibit unexpected behavior.

• **Ethical challenges:** Use of live biological materials would provoke some ethical arguments.

• **Contamination risks:** Biological components hold a potential of harming environment.

Artificial biological intelligence offers a new calculation paradigm inspired by nature, and going beyond imitating the cognitive capacity of the human brain. Having the technology progressed this way, will blur the boundaries between artificial and biological worlds.

Universal Intuition

Universal intuition, describes foresight abilities which are beyond human perception, and based on data, but it is difficult to explain it through a classical logic. This concept represents a phase where systems could make intuitive estimations with the integration of AI, quantum calculations and big data analyses.

How could universal intuition be possible?

Universal intuition could be possible through the following technological and scientific developments beyond statistical analyses:

• **Quantum Artificial Intelligence (QAI):** Simultaneously processing millions of possibilities, QAI could offer non-deterministic and creative solutions.

• **Big data and deep learning:** It can make intuitive-like predictions through exploring patterns which cannot be perceived by the human eye or mind.

• **Collective consciousness network:** The networks connecting human brains and machine could create a joint "consciousness" level, and this consciousness would be able to develop intuition.

- **Chaos theory:** At the complex systems, it would be possible to make predictions for the events in the future by figuring out the fundamental orders of the seemingly accidental happenings.

Fields of application

- **Forecast of global crisis:** The early predictions of large-scale crises such as climate change, financial collapse or epidemics could be possible.

- **Economic and social dynamics:** Economic collapse or political rebellion could be intuitively predicted beyond existing data.

- **Health and medicine:** The disease risks which are not yet noticed by the individual could be perceived "intuitively" and the preventative treatment processes could be started.

- **Creative problem solving:** Solutions which are not possible by classical methods in science and engineering could be discovered.

Advantages

- **Increase of prediction capacity:** It enables the early intervention opportunity for the events to happen in the future.

- **Reducing risk:** The likelihood of noticing the potential threats in the chaotic systems early enough could be increased.

- **Enhancing creativity:** Solutions beyond human mind could be figured out.

Challenges

- **Ethical issues:** Use of the systems which predict the future, might threaten human freedom.

- **Misleading predictions:** Misleading intuition migh come from too much complex and unpredictable systems.

- **Security threats:** These technologies could be manipulated by evil-minded actors.

Universal intuitive represents a new intuitive level which could expand the boundaries of human mind, and shed light on the unknown. But this new development will bring up new subjects on the questions of etchics, security and philosophy.

Autonomous Scientific Discovery Systems

Through the integration of AI and robotic technologies, autonomous scientific discovery systems design experiments by themselves, refining hypoteses, and conducting scientific research. These systems could make scientific strides without requiring human intervention.

What makes it work?

1. Data analysis: It identifies the hidden patterns and relations inside big data groups.

2. Hypothesizing: It develops likely scientific hypotheses through statistical and logical inferences.

3. Design of the experiment: It designates the experiment conditions and generates appropriate protocols for experiments.

4. Realizationf of the experiment: It conducts experiments in physical and virtual settings at robotic labs.

5. Analysis of the results: It evaluates the results of the experiment and reports the successful discoveries.

Prospective applications

• **Discovery of drugs and medicine:** By analyzing the data related to human biology, it can discover new drugs. Additionaly, it can speed up the clinical experiment phases.

• **Materials science:** New and strong materials (for example super conducting materials)

• **Physics and Cosmology:** It can make comlex simulations in order to make out better physical laws in the universe.

• **Environmental science:** By analyzing ecological systems, it can come up with sustainable solutions.

- **Genetic and biology:** It can conduct autonomous experiments for GENOM arrangements and discovery of new biological processes.

Advantages

- **Speed:** It can conduct experiments and make data anaylysis quicker.

- **Creativity:** It would recommend hypotheses which went unnoticed by humans

- **Low cost:** It can lower the costs in the long run.

- **Uninterrupted work:** With the capacity of 7/24 working hours, it can speed up the scientific discovery process.

Challenges

- **Data bias:** By receiving wrong and incomplete data, it can end up having inaccurate hypotheses.

- **Ethical issues:** Some risky research projects such as genetic experiments could be out of control.

- **The role of scientists as a human being:** Having no human involvement in the scientific process could lead to etchical and philosophical discussions.

- **Security risks:** Espeacially the risk of misuse could be the case in the fields of biotechnology and AI.

Autonomous scientific discovery systems could offer new solutions to many seeminly unsolved problems by speeding up the scientific revoluitions in the history of humankind. But it is vitally important to manage these systems in a right way and maintain them within the ethical boundaries.

THE EVOLUTION OF THE DEFINITION OF HUMAN

Combination of biological and digital systems could influence deeply the fundamental structure of the human experiment. While these developments have been shaping the definitions of self concept, and human on the one hand, new ethical issues could emerge. Building up human identity, counciousness and free will on the digital and biological foundations would throw a serious of problems which would pusch the limits of ethics into question.

Change in the definition of self

1. Combination of biological and digital identity

Shift of self-perception: Technologies such as brain-computer interfaces, AI integration or digital consciousness uploading could expand or change the sense of self of individuals. Humans could maintain their existence as digital entities or they could leave their biological bodies.

• **Digital identity:** Humans can exist in the digital worls as virtual beings (avatars), yet this situation could alter their perception of identity. To what degree the digital sense of being should match up with the physical sense of being?

2. Consciousness and personality modification

• Humans could change or enhance their consciousness in order to make themselves more productive. This could cause serious problems on the free will and personality:

- Could the indiviuals who changed their identity be considered still the same person?

- How would these kinds of changes affect individual freedom and psychological balance?

New human definition: interaction between AI and humanity

3. AI and human limits

- Integration of the human brain with AI could surpass the cognitive abilities of humans, thus making it "superior" in a sense, becoming a threat to the definition of humanity.

- Where should the boundary between human and AI be drawn?

- Should AI be treated like a human when AI developed a human-like consciousness?

4. Hybrid humans

- Genetic makeup, biological organs and minds of humans could be united with digital or artificial components. How could these kinds of "hybrid" humans adapt themselves to the traditional definition of being a human?

- Would humanity lose its its biological existence?

- Would technological modifications inhibit humans from evolutionary potential?

Challenges

5. Free will and manipulation

Brain-computer interfaces and emotional manipulation: Direct intervention in human brain could restrict free will of individuals.

- To what point could these types of interventions be acceptable?

- How could manipulation have an impact on the social order, beyond establishing control over humans?

6. Privacy and identiy security

Leak of digital identities and use of personal data: The fact that access to digital senses of being and personal information becoming widespread could increase the risk of identity theft and invasion of privacy.

• How should the rights of humans on their digital sense of self be protected?

• How should the theft and abuse of digital sense of self be handled etchically?

7. Human-animal (hybrid) living being interactions

• Advancements in genetic engineering and biotechnology could remove the borders between humans and animals.

• Could genetic modifications featuring humanlike changes be accepted ethically?

• How should the rights of "Hybird" entities be defined?

8. Social justice and inequality of access

Access to technology and social justice: Technologies increasing mental capacity could deepen even further the inequality in the core areas such as education and health.

• Inequality to technology could lead conflicts between new social classes.

• How the technologies should be distributed gains importance in terms of reaping social benefits.

9. Evolution of the definition of humanity and rights

• Technological and biological changes can transform the definition of human.

• What does being a "human" mean?

• Should the artificial consciousness or hybrid beings be included withing the concept of human rights?

• How should the biological and digital right of humanity be defined?

These evolutionary changes in the definition of the sense of self would complicate etchical issues. Humanity would not only be defined by biological limits, but also redefined by digital and AI integrations. This process would necessitate reshaping the fundamental ethical values such as freedom, identity, privacy, and equality. The solve these questions both on an individual and social levels shall determine the way technological developments would be shaping humanity in the future.

DOMINATION OF VIRTUAL LIFE

Domination of virtual life is a process which obscures the boundaries of real life with digital and virtual worlds. This period, where humans start spending more time with business, entertainment, social interactions, and personal develeopment in their day to day life, could cause socially, culturally and ethically deep transformations.

Areas to be domitated by virtual life

1. Social interactions and communications

Virtual Reality (VR) and Augmented Reality (AR)

• Virtual reality and augmented reality, enable humans to interact socially without having to come together physically. This could make it possible for humans to forge deep connections with their friends, families and co-workers.

• Humans can conduct social experiences in virtual world through digital avatars, which are impossible in real world settings.

2. Business and education world

Digital offices and remote work

• Remote work which became prevalent in post pandemic period, ensured that virtual platforms got a permanent place in business world. Enhanced virtual meetings, joint working spaces and tools of digital interactions might start replacing physical offices.

• Virtual schools and digital courses as well as interactive educational environments in education might become widespread. Students from all over the world could follow the lessons by joining the virtual classes.

3. Entertainment and the media

Virtual worlds and video games

• Virtual worlds such as started getting an important place in the entertainment industry. People could organize social events, play games, attend concerts or create digital art works in virtual worlds.

• These types of platforms could support digital economy, and enable individuals to go through different cultural experiences via virtual reality.

4. Personal development and management of life

Virtual therapy and health services

• Virtual environments could integrate with pshychological and physical health services. Online therapy, digital exercise programs, and virtual meditation sessions could create a new avenue for personal development.

• Individuals could express themselves freer, and experience different identities and lifestyles in the virtual world.

Advantages

Access and Equality

• By clearing the way of the geographical, economical and physical obstacles, virtual platforms could offer opportunities to many more people. Peopla could have educational opportunites from faraway places, fing jobs and attend social activities.

Freedom of personal expression

• People could assume any identiy in the virtual world, exhibit different personalities there and express themselves independently free from social norms.

Mental and emotional discovery

- By offering experiences on different reality levels, virtual life enables them to make emotional and mental discoveries. This could also be used as a therapeutic tool.

Challenges

1. Issues of identity and reality1

Blurring the border between virtual and real

As the people spend more time in the virtual world, they may face risk of breaking from reality. Spending less time in the physical world, and "living" in the virtual world, could lead to some psyhocologial health problems.

- Virtua identities cause individuals to lose themselves or to go through identity crises.

2. Social loneliness and isolation

Insufficiency of digital social connections

- Virtual interactions may not substitute for real, face-to-face relationships. While connectiong with one another in the digital world, people could be deprived of deeper emotional and physical interactions.

- Indivituals who spent time in virtual life for too long, may face feelings of loneliness and isolation.

3. Addiction and loss of control

Spending too much time in the virtual world

- Increased time spent in the virtual world could create addiction inducing effects. By focusing too much on the satisfying experiences offered by the virtual world, people could neglect their responsibilities in the physical world.

- This situation could cause people to be negligent of their relationships and duties in the real world.

4. Problems of Ethics and Security

Data security and privacy

• Virtual world could carry serious risks as far as collecting, monitoring, and abusing of personal information are concerned. Activities of the individuals in the virtual worlds could be manipulated through the data collected for commercial purposes.

Manipulation and guided experiences

• Dijital platform could make use of algorithms to guide users in the direction of certain commercial interests and to affect them. This could make the free will of individuals questionable.

Domination of virtual life could transform the social, psychological and cultural make-up of the individuals to a great extend. The fact that people spend much more time in the virtual worlds could offer too many opportunities, but at the same time, it might bring psychological, ethical and social problems. Integration of the virtual worlds with the physical world in a balanced way would be the key to creating a healty and sustainable lifestyle in the new era in the future.

HUMAN EXPERIMENT OF THE FUTURE

Expansion of the emotional experiences means enhancement of the feeling capacity of humans, and the discovery of deeper, complex and new emotings beyond the existing ones. This development could only be possible through both biotechnological innovations and increase of mental awareness.

Alternative ways and technologies

1. Integration of brain-machine

• Thanks to brain-computer interfaces, human mind could be programmed to accommodate new emotional states.

• Simulated emotions: Denser and more personalized emotional experiences could be created in the virtual reality environments.

2. Empaty development through AI

• Personal AI assistants could analyze the emotional states of the users to give them an appropriate emotional feedback.

• Social algorithm which enhances empathy, could create deeper connections in human communication.

3. Neurotransmitter substance modulation

• Direct control of neurotransmitter substances such as dopamine, serotonin, and oxytocin, could lead to management of emotional experiences.

• Through designing personalized "happiness regime", chronic stree or emotional lethargy could be prevented.

4. Emotion based virtual environments

• Platforms such as metaverse could enable people to go through dense emotional simulations they never experienced before.

• They can make emotional experiences which are impossible in real world to be lived digitally.

5. Discovery of new emotions

• With the increase of the capacity of the human brain, new type of emotional reactions including collective happiness or multi-layered empathy could thrive.

• By having an expanded emotional spectrum, and not limited with the fundamental feelings such as happiness, fear and rage, a richer emotional world could emerge.

Advantages

• **Getting human connections strong:** Deeper empathy and understanding could enrich interpersonal relations.

• **Psychological healing:** Overcoming traumas quickly and supporting mental health could be possible.

• **Personal development:** New emotional experiences could make individuals more creative and adaptable.

• **Artistic and creative potential:** More sophisticated and dense emotional expressions in arts and literature could emerge.

Challenges

• **Emotional excess:** Experiences such as excessive happiness or intense sadness could lead to psychological imbalances.

• **Addiction:** Through chemical or digital manipulations, addiction to artificial emotions could intensify.

• **Distorted perception of reality:** The boundary between digital emotional simulations and real world would become blurry.

• **Manipulation:** Being able to control emotions could pave the way to manipulating individuals psychologically.

Expansion of emotional experiences could bring deeper meaning and richness to human life. But this development should be kept carefully within the ethical boundaries. Protecting human emotions and developing them within healty boundaries have critical role to play in this transformation.

NEW MIND SPACE

Collective consciousness networks indicate a consciousness level where knowledge and experiences humans and machines generate together are shared. This network creates a joint counsciosness by merging the data coming from human brains, AI systems, even from biological and digital platforms. These structure could make societies and technological systems exist side by side close to each other in a harmonious way.

What makes it work?

1. Brain-computer integration

• Information transfer from the brains of the humans is possible. Thanks to the brain-computer interfaces, individuals could transfer their thoughts and emotional states to one another.

• Humans can share intellectual processes and experiences in a digital platform with others.

2. AI and data sharing

• By analyzing the thoughts of individuals, AI can gather this data in a collective pool.

• Algorithms collect data related to the joint consciousness by monitoring the mental processes of human brains and their reactions.

3. Social connections and communications

• By emergin with social consciousness, personal information networks, could create a wider social mind.

• Forging momentary emotional and intellectual ties, would increase the capacity of collective thinking and decision making

4. Ditital and biological integration

• Technology integrates biological systems and combines the biological and digital information of each individual.

• People can make collective decisions by sharing each other's thoughts and ideas in a digital environment.

Applications

1. Global decision making and crisis managament

• Humans and AI could analyze social problems and develop policies faster within the framework of joint consciousness, and solve global crises in an effective way.

• Collective consciousness could come up with solutions to the problems such as climate change, management of pandemic, and war.

2. Education and learning systems

• Thanks to the collective consciousness, students could have an access to information worldwide, and speed up their learning processes.

• They can offer more personalized learning methods by analyzing the individuals' command of knowledge.

3. Arts and creativity

• Collective consciousness can ensure that ideas coming from different cultures to be combined and lead to emergence of new artistic expressions and creative projects.

• It can create instant and more effective possibilities of cooperation among artists and scientists.

4. Health and psychology

• Mental disorders and psychological problems could be diagnosed through collective consciousness and treated fast.

• The mental and emotional states of the people could be cured by establishing connections with others.

Advantages

• **Fast and effective decision making:** The collective consciousness network, could help making fast and right decisions beyond the limits of the human mind.

• **Deeper empathy and understanding:** Access to the thoughts and emotional states of different individuals could enhance empathy and social ties.

• **Global information sharing:** By having a common pool of knowledge, humanity could function more productive and effective.

• **Creative innovation:** Collective consciousness could elevate the cooperation between human and AI up to the highest level in order to develop new and creative solutions.

Challenges

• **Loss of individual identity:** Collective consciousness network could threaten individual thought and identity. In order to comply with the social pressures, people could lose their own original thoughts.

• **Data security and privacy:** Sharing thoughts and feelings of people on digital platforms could cause a huge privacy and security problem.

• **Loss of control:** If collective consciousness has a central structure, then the risk of misuse of this network might occur.

• **Philosophical problems:** Humanity could question the ethical and philosophical dimensions of connecting with collective consciousness. Changes may occur in the processes of personal freedom, thought and decision making.

Collective consciousness networks could be a revolutionary step for humanity. However, developing this technology should be managed carefully with respect to ethics, security and personal freedoms. These networks could strengthen the social ties and reshape the human experience. But at the same time, a solid control is necessary in order to protect individual freedom, identiy and privacy.

HUMAN AND AI INTEGRATION

Unlimited increase in mental capacity means that by surpassing the biological limits of human brain, processing power, memory capacity, and learning speed could get to the indefinite level. This concept coul be materialized through both biological advancements and intelligence-enabled technology

Technological and scientific approaches

1. Brain-computer interfaces, BCI)

• Initiatives such as Elon Musk's Neuralink projects aim at integrating the brain directly with the computers.

• Thanks to these systems, accessing information instantly, storing memory in external stroges, and controlling devices through mental command will be possible.

2. Artificial intelligence supported brain

• Intelligence modules integrated in the brain could increase the information processing capacity.

• In solving the complex problems, it could be possible to surpass the limits of human brain.

3. Nanotetechnology and neuron development

• Nano robots embedded in the brain could repair neurons and accelerate their processing speed.

• By optimizing neural connections, information transfer speed could be increased.

4. Genetic manipulation

• Through gene regulation techniques such as CRISPR, genetic changes which would optimize learning and memory processes could be made.

5. Quantum computing integration

• Integration of brain and quantum processers can increase computational speed and capacity to unlimited levels.

6. Consciousness backup and digitilization

• By "uploading" to external systems, consciousness can be free from the physical limits of the brain.

• Human mind could interact in digital platforms rather than in biological body.

Advantages

• **Improving memory:** The human brain can store almost unlimited data.

• **Instant access to information:** It can be possible to access information directly without connecting to external sources such as internet.

• **Creative problem solving:** People can solve complex scientific and technological problems much faster.

• **Accelerated learning:** It may be possible to master a subject in minutes instead of years of training.

Challenges

• **Problem of identity and sense of self:** Excessive expansion of mental capacity can distort an individual's sense of identity.

• **Overload:** Information overload could lead to mental health problems.

• **Ethical issues:** Manipulation of the human mind and cloning of digital selves may become controversial.

• **Inequality:** The fact that only certain segments have access to this technology could lead to new social class divisions.

Unlimited increase in mental capacity could start a revolutionary period in the evolutionary process of humanity. However, its is of great importance that this technology is managed in a right way, ethical boundaries are determined and human nature is protected.

DIGITAL IMMORTALITY

Death has been one of the greatest mysteries in human history. However, with the advancement of digital technologies and biotechnologies, today the boundaries of death are being questioned again. Digital immortality refers to the idea that an individual's consciousness or personality can survive physical death by being preserved in a digital environment. This concept can profoundly affect the meaning of death and life, but also raises many ethical, philosophical and social issues.

1. Consciousness uploading (*Mind Uploading*)

• By converting the structure and function of the brain into a digital format, a person's thoughts, memories and personality are transferred to a computer environment. This technology can provide the opportunity to live a life independent of the biological body.

• Human consciousness may continue to live on as an artificial intelligence, but is that person really the "same" person? The answer to this question forms the philosophical foundations of digital immortality.

2. Virtual avatars

• A digital avatar of a person can be created in virtual reality or augmented reality platforms. This avatar enables the person to maintain social relationships, interact, and continue living in the digital environment.

- However, this avatar doesn't reflect the individual's true thoughts and feelings, it would only mimics pre-recorded information and personal style.

3. Digital memory and storage of data

- It may be possible to reconstruct a person's life in a virtual fashion by bringing together the digital memories (photos, videos, messages) that a person has created throughout his or her life.

- But this leaves a kind of "digital legacy of the person, and deep questions about his or her personality of consciousness remain unanswered.

Advantages

- **Possibility of immortality:** Humans may continue to live as a sort of "digital being" even after biological death. This means taking one step beyond the oldest desire in human history. Digital immortality offers opportunity to preserve a person's cultural heritage, ideas and ideal for generations to come.

- **Protection of knowledge and experience:** By storing invididuals's entire lives in a digital environment, knowledge, experience and wisdom can be passed down through generations. This could create a revolutionary knowledge pool for humanity. People can interact with their own digital presence of that of past thinkers and benefit from their ideas.

- **Personal development and eternal existence:** Digital immortaliy could enable a person to continue his or her personal development. People can gain new experiences by improving their presence in digital form. This may raise new questions about what it would feel like psychologically to be an immortal being.

Challenges

- **Identity and problem of sense of self:** The most fundamental question regarding digital immortality is whether an individual's identity can be preserved in digital form. If a complete digital backup of the brain is made, can this "new entity" still be considered the same person? Transferring one's thoughts, feelings and memories to digital media may not mean a complete "re-creation" of the personality. Are we the same individual, or just and AI similar to him or her?

• **Ethical and privacy issues:** People's digital assets can create a majör privacy problem because they contain people's perfonal information, thoughts and emotions. Digital immortality can be misused by others without that person's consent. Additionally, the continued existence of a person's digital presence after his or her death may be contrary to that person's true will. Family members or kinspeople may object to the maintenance of digital assets after death.

• **Social and psychological effects:** Digital immortality can profoundly affect social structures and human relations. People can interact with immortal digital beings and have difficulty coming to terms with the reality of death. Moreover, social views may vary as to whether death should be accepted as a natural process or not. Digital immortality could radically change people's perceptions of death.

• **Artificial intelligence and human rights:** If a person "lives" in a digital environment, should he or she have these existence rights? Digital immortality could create a new field regarding human rights and ethical values. There can be a deep debate about whether artificial intelligence should have the rights and freedoms of individuals.

Digital immortality and the future of humanity

Digital immortality could radically change humanity's thoughts about life and death. Though this may seem like science fiction, it has the potential to become reality with technological advances. However, such an idea of immortality may bring many difficulties in terms of human identity, ethical values and social structures. The end of death and digital immortality is not only a scientific issue, but also a philosophical and ethical challenge. As humanity seeks answers to these questions in the future, it will develop a new understanding of how to establish the balane between life and death.

II

DIGITILIZATION OF THE BODY AND SCIENTIFIC BREAKTHROUGHS

DIGITAL EVOLUTION OF THE HUMAN BODY

Digitalization of the body is a concept that arises from the combination of biological and dijital technologies. This process involves the integration of the biological body with digital sistems, monitoring and controlling people's biological functions through digital platforms. The digitalization of the body could create a major transformation in many areas including health, education, work force, social interactions, and personal experiences. The digitalization of the human body has the potential to create a wide range of impact from medicine to our daily lives.

Key components of the digitalization of the body

Biometric data monitoring and collection

One of the most common applications of body digitalization is the continuous monitoring of biometric data. The body can continuously transfer data to the digital environment through wearable devices, implants of subcutaneous sensors.

- **Biometric sensors:** Smartwatches, wearable devices, or clothing with sensors can monitor heart rate, blood pressure, oxygen levels, body temperature, and other biological parameters in real time. These devices can monitor health status, providing early diagnosis, treatment, and health management.

- **Genetic monitoring:** Biological data collection can be combined with DNA analysis and genetic engineering. Genetic information can be stored on digital platforms and deeper analysis can be done to monitor health. For instance, it may be possible to detect genetic predispositions or predict the development of diseases.

Digital identity and virtual reflection

The digitalization of the body can make it possible to create a reflection of the physical being in a virtual environment. People can exist in the virtual world through digital identities or avatars.

• **Digital identity:** People can use their biometric data as digital identity. For example, digital identity verification can be done with fingerprints, retina scans or DNA profiles. This enables us to securely manage all types of data from health data to our personal information in a digital environment.

• *Digitizing their bodies offers a more interactive and* realistic experience in the virtual world

• **Virtuality and avatars:** People can exist with digital avatars in virtual reality (VR) or augmented reality (AR) environments. Digitizing their bodies offers a more interactive and realistic experience in the virtual world. This process allows for greater interaction on social media, games, or virtual meetings.

• **Brain-computer interfaces (BCI)**

Brain-computer interfaces connect brain signals directly to digital devices. This builds a bridge between the physical and digital worlds.

• **Mental control and interaction:** hanks to BCI technology, people can control computers or digital devices with mental commands. This could improve the quality of life for paralyzed patients or allow a person to transmit thoughts directly into a computer system.

• **Mind upgrade and digitization:** The integration of brain functions with digital environments may allow people to expand their cognitive abilities in the digital environment. For example, it is possible to record a person's thought processes, store this data, and replay it or access it via digital platforms.

Body implants and nanotechnologies

Digitalization of the body involves profound biological interventions through implants and nanotechnologies.

• **Nano-implants and smart prosthetics:** Nanotechnologies can digitize the body with micro-level devices integrated into the body. These devices can monitor, repair or improve biological functions in the body. For example, nano-implants could be used to continuously monitor digital health data and directly intervene in a person's metabolism.

• **Genetic modification and biotechnological upgrades:** Genetic engineering can change the biological characteristics of humans. Improvements in human biological structures through digitalization may include situations such as curing genetic diseases or delaying biological aging.

Digital upgrade and increased performance

Digitized bodies can have various digital and biotechnological upgrades to increase their biological capabilities.

• **Fast data processing and augmented intelligence:** People can have the capacity to think and process information faster through brain-computer interfaces supported by artificial intelligence and digital technologies. This can speed up decision-making processes and improve cognitive abilities, especially in professional life.

• **Increased physical performance:** Digitized bodies can enhance their physical performance using biotechnological implants and smart devices. For example, smart prosthetics, bionic limbs, and support devices can give the body physical strength and endurance.

Potential applications of the digitalization of the body

Health and medicine

The digitalization of the body could revolutionize the healthcare industry. Applications such as real-time health monitoring, early diagnosis, personalized treatment, and genetic engineering could shape the future of medicine.

- **Personalized medicine:** Digital body monitoring could make it possible to create personalized health plans for each individual. By combining biometric data, genetic analysis, and environmental factors, specific treatment methods could be developed for each individual.

- **Remote monitoring and treatment:** Digitized bodies could enable healthcare professionals to continuously monitor their patients. For example, management of chronic conditions such as heart disease or diabetes could become more efficient thanks to continuous monitoring devices.

Education and workforce

Digitized bodies could revolutionize education and the workforce by enhancing people's cognitive and physical abilities.

- **Digital education:** Thanks to smart devices and brain-computer interfaces, people can learn faster and develop their skills in digital environments. Educational processes can be adapted to individual needs.

- **Increased workforce efficiency:** Digitalized bodies in work environments can increase people's workforce. High-performing employees are supported by digital technologies and become faster and more effective in their work processes.

Social and cultural change

The digitalization of the body can have profound effects on social relations, identity, and social interactions.

- **Social interaction and identity:** Digital identities can change the way people interact socially. With technologies such as virtual reality and augmented reality, people can interact more through virtual environments. People can experience each other more interactively in virtual environments through digital avatars.

- **New social structures:** Digitized bodies can change social structures. Physical and biological differences between people can diminish, as digital upgrades can help build an egalitarian society. However, this transformation can also raise issues of social class, discrimination, and ethics.

- **Security issues**

The digitalization of the body may raise important ethical and security concerns.

- **Privacy and data security:** Storing and processing biometric data in a digital environment may bring about problems such as privacy violation and cyber attacks.

- **Biotechnological injustices and discrimination:** Digitized bodies may increase social inequality as some individuals gain easier access to biological and digital enhancements.

- **Biological rights and identity:** The digitalization of the human body can raise questions of individual rights, free will, and identity. The control and manipulation of digitalized bodies can push ethical boundaries.

- The digitization of the body is a technological advancement that could radically transform the human experience. Combining both physical and cognitive capabilities with digital platforms could make people healthier, stronger, and more productive.

QUANTUM

Quantum artificial intelligence (QAI) is a concept that combines classical AI with quantum computing. Quantum computers, unlike classical computers, can perform multiple operations simultaneously thanks to quantum bits (qubits). This could dramatically improve the learning, prediction and optimization capabilities of AI.

How does it work?

The main difference of quantum artificial intelligence is that it is based on the guidelines and principles of quantum mechanics.

• **Superposition:** Qubits can be in both 0 and 1 states simultaneously, providing the ability to consider many possibilities simultaneously.

• **Entanglement:** Other qubits associated with a qubit change simultaneously, even if they are far apart. This greatly increases data processing capacity.

• **Quantum tunneling:** Algorithms can find faster solutions by overcoming classical obstacles.

Quantum AI applications

• **Optimization problems:** Complex problems such as logistics, energy distribution and financial portfolio management can be solved in seconds.

• **Drug development and moleculer similation:** Quantum AI could speed up drug development by simulating biological molecules.

- **Encryption and cyber security:** Systems strong enough to break classical encryption methods can be developed. At the same time, more secure quantum encryption methods may become possible.

- **Complex data analysis:** Quantum artificial intelligence can overcome the limitations of classical methods in big data analysis and make more accurate predictions.

- **Artificial general intelligence (AGI) development:** Quantum processing power could pave the way for the concept of artificial general intelligence (AGI), which would enable AI to think and learn like humans.

Risks

- **Control problem:** If mismanaged, quantum AI could create chaos in economic, military, and social systems.

- **Cyber threats:** Existing encryption systems can be easily broken.

- **Ethical issues:** The complete automation of human decision-making processes could lead to new ethical debates.

Quantum artificial intelligence has revolutionary potential in the field of science and technology. However, the management of this technology and the determination of its ethical limits will be of vital importance for humanity. Possible developments that may push the boundaries of reason and statistical science in the future may deeply affect human history. Here are some striking headlines:

1. Quantum AI (QAI)

Quantum computers have a data processing capacity far beyond that of classical computers. This capacity will enable artificial intelligence to produce solutions to more creative and unpredictable problems. Even statistical analysis methods may not be able to keep up with such data density.

2. Sentient (emotional and conscious) algorithms

It is expected that systems will emerge where artificial intelligence will not only be data-driven but will also be able to make decisions based on emotion, ethics and intuition. This may require redefining the concepts of reason and logic.

3. Artificial biological intelligence

Biology and statistics will come together thanks to organism-based computers (biological processors). These systems can imitate the thought processes of the human brain, based on both biological and digital information.

4. Universal intuition models

Statistics are usually based on historical data. However, in the future, with instantaneous data flow and artificial intelligence, predictions will be made based on "hunch" models rather than probabilities.

5. Autonomous scientific exploration systems

Artificial intelligence will be able to develop and test its own scientific hypotheses, enabling a constantly evolving knowledge system without human intervention.

6. Human-machine consciousness unification

Thanks to neural connections, the human brain may be able to perform direct statistical analysis. This may reveal a new dimension of intelligence where individuals blend intuition and data analysis. These developments will necessitate not only technical advances but also ethical, social and cultural transformations for humanity.

The fusion of man, machine and consciousness ("cybernetic fusion" or "transhumanism") could be one of the most radical transformations of the future. Here are the possible consequences of this fusion:

1. Unlimited increase in mental capacaity

• When the brain is directly connected to the internet, instant access to information will be possible.

• Memory, attention and calculation skills can become supported by artificial intelligence. People can become beings that "do not forget".

2. Expansion of emotional experiences

• By intervening in the chemical structure of the brain, mental states such as stress and depression can be easily regulated.

• Even "new emotions" of a type that people have never experienced can be created.

3. Physiological and physical transformation

• Physical disabilities can become a thing of the past. Artificial limbs, organs and nerve connections can strengthen the human body.

• The aging process can be stopped or reversed thanks to advanced health technologies.

4. Collective consciousness networks

• People can connect their minds and share thoughts directly.

• This could change the concept of individual consciousness and lead to new forms of social organization.

5. New ethical issues

• If consciousness transfer becomes possible, the definition of "self" and "human" will change.

• Security issues such as the risk of hacking the human mind or privacy of thought may arise.

6. The dominance of virtual life

• It may become common to disconnect from physical reality and live in virtual worlds that are experienced entirely mentally.

• Work, education, entertainment, and social interactions can all occur on a mental level.

7. The end of death and "digital immortality"

• When consciousness can be transferred to digital systems, the concept of death may lose its meaning.

• Even if people lose their physical bodies, they can continue to live as digital beings.

8. Human-machine class distinction

• Great social inequalities may arise between those who have access to these technologies and those who do not.

• This situation may cause conflicts between the "biological human" and "enhanced human" species.

CELLULAR REPAIR WITH NANOTECHNOLOGY

Nanotechnology is a branch of science that deals with the manipulation of materials and devices at the atomic or molecular level. This technology has the potential to revolutionize medicine and biology. Cellular repair refers to the repair of damaged cells or tissues in the body, and nanotechnology can play a critical role in this process. The use of nanotechnology in cellular repair opens up a promising area for the solution of incurable diseases and organ damage.

Fundamental principles of nanotechnology and cellular repair

Nanotechnology makes it possible to perform interventions at the cellular level by using technologies such as bioengineering, biomaterials, nano-machines and molecular machines. Nanotechnological devices and materials can directly affect the cells without damaging them and thus initiate healing, regeneration and repair processes.

1. Nanorobots and nanomachines

• **Nanorobots:** Nanorobots produced with nanotechnology are very small devices that can work at the cellular level. After being injected into the body, these robots can reach targeted areas through the bloodstream and perform repairs at the cellular level.

• **Cellular repair:** Nanorobots can detect damaged cells, repair them or generate new ones. For example, they can detect and destroy cancer cells or repair heart muscle damage.

• **Drug distribution:** Nanorobots can be used to deliver targeted therapeutic agents to the correct cells, thereby minimizing side effects.

2. Nanoparticles and drug delivery systems

• **Nanoparticles:** Nanoparticles enable drugs to be delivered more effectively into the body. These particles can pass through cell membranes and deliver drugs directly to damaged cells. This way, the treatment only works on the targeted cells, leaving healthy cells unharmed.

• **Cellular repair:** Nanoparticles can carry drugs, gene therapies, or growth factors harmlessly to cells. This can heal tissue damage, promote cell regeneration, and restore organ function.

3. Genetik modification ve nanoteknoloji

• **Gene therapy and CRISPR technology**: Nanotechnology can make genetic engineering applications more efficient. For example, genetic editing techniques such as CRISPR can be delivered into cells using nanoparticles, allowing damaged or diseased genes to be repaired.

• **Cellular repair**: Nanotechnology can prevent diseases or repair cellular damage by making genetic adjustments. For example, in cases of muscle diseases or heart disease, these methods can regenerate cells or correct the genetic roots of diseases.

Application areas of cellular repair with nanotechnology

1. Cancer treatment and cellular repair

• In cancer treatment, cancer cells can be targeted and destroyed in very small sizes thanks to nanotechnology. Nanoparticles can treat tumors more effectively by carrying drugs to cancer cells. Nanorobots can directly target cancer cells.

• In addition, cellular repair can be provided to repair tissue damage that occurs after cancer treatment and to protect healthy cells.

2. Cardiovascular diseases and heart regeneration

• When heart muscle cells are damaged, the ability of heart tissue to heal is limited. Nanotechnology can treat damaged heart cells with gene therapy or growth factors, allowing heart tissue to regenerate.

• Nanotechnology can also encourage the proliferation of heart muscle cells and prevent loss of function after a heart attack.

3. Neurological diseases and brain regeneration

• The brain and nervous system have limited capacity for cellular repair. Nanotechnology could revolutionize the treatment of neurological diseases. Nanorobots and nanoparticles could enable drugs to reach their target directly without damaging nerve cells.

• Nanotechnological solutions for cellular repair have potential, especially in the treatment of Parkinson's disease, Alzheimer's and spinal cord injuries. Nanotechnology could be used to repair brain cells, reconnect damaged neural networks or create new ones.

4. Anemia and bone repair

• In the treatment of diseases such as bone damage or osteoporosis, nanotechnology can encourage the regrowth of bone cells. Nanoparticles can accelerate the proliferation of bone cells and provide renewal of bone tissue.

• In addition, nanotechnological materials can be used to make bone tissue more durable and to heal bone fractures faster.

5. Advantages and challenges of cellular repair with nanotechnology

Advantages

- **High targeting precision:** Nanotechnology can increase the effect of treatment and minimize side effects by delivering treatment agents directly to targeted cells or tissues.

- **More effective treatment:** Repair at the cellular level can help achieve more effective results in the treatment of diseases. For example, it can be an important step in the treatment of chronic diseases such as cancer or heart disease.

- **Comprehensive repair:** Nanotechnological devices can provide holistic healing of the body by performing repairs at both the genetic and cellular levels.

Challenges

- **Safety and side effects:** Since nanotechnology intervenes in biological systems, its long-term effects need to be fully understood. Accumulation of nanoparticles in the body or misdirection can lead to serious side effects.

- **Ethical and societal issues:** Genetic interventions and cellular repairs can raise ethical issues. The societal impacts of modifications to human cells can deepen ethical debates about genetic engineering and biotechnology.

- **Technological challenges:** The integration of nanotechnology into clinical applications faces numerous technical hurdles. More research is needed to ensure that nanodevices are properly injected into the body, reach target cells, and operate safely.

6. The future of nanotechnology in cellular repair

Nanotechnology has great potential in cellular repair and treatment processes. It can be an effective solution in many areas, from cancer treatment to heart regeneration, from neurological disease treatment to bone repair. However, more research, safety tests and ethical discussions are needed for this technology to be put into clinical applications. In the future, repairs made at the cellular level in the body with nanotechnology may be one of the most important revolutions in medicine.

PRODUCTION OF BIOLOGICAL ORGANS

Reproduction of biological organs is one of the most exciting and potentially revolutionary areas of medicine. The human body may experience organ loss over the years due to various diseases, injuries or aging. However, biotechnological developments can radically transform organ transplantation and treatment processes. The ultimate goal in this field is to regenerate or repair lost or damaged organs in the human body.

Regeneration of biological organs: Basic technologies

Cell reprogramming and genetic modification

- **Induced Pluripotent Stem Cells (IPS cells)**

IPS cells are a technique to recycle a mature cell into a stem cell. These cells can develop into all body cells and have the potential to regenerate various organs.

For organ regeneration, these cells can be programmed to form specific organs by directing them in the right way. This can form the basis for the production of organs such as the heart, liver or kidney.

Also, genetic modification techniques can be used to make organs more resistant to diseases or to slow down aging.

3D biyoyazıcılar ve organ baskısı

- **Bioprinters (Bioprinting):** 3D bioprinters are devices

used to build organ tissues. These devices can create biological structures, blood vessels, and other organ structures by arranging cells in a specific order.

Organs can be "printed" layer by layer using cells and biomaterials. This technology can make organs not only functional but also biocompatible.

While skin, cartilage, and some tissues are currently the most commonly produced, organ-level production may be possible in the coming years.

Organ augmentation and reconstruction

• **Tissue engineering:** Another way to regenerate organs is to use tissue engineering to repair or improve existing organs. This means using bioengineering techniques to replace the missing part of an organ in the body and allow it to heal. Organ tissue can be produced using biological matrices, cell cultures, and growth factors. For example, such methods can be used to repair damaged heart tissue or to prevent immune rejection after organ transplantation.

Organ transplantation and paired organs   (Xenotransplantation)

• **Xenotransplantation:** This method is a field of research that focuses on the use of animal organs on humans. Through genetic engineering, organs obtained from pigs, for example, can be transplanted into humans. However, organ rejection, immune responses and ethical issues are still important obstacles to be solved in this area.

Organ regrowth and regeneration

• Organ regeneration: Some organs, such as the liver and skin, can naturally regenerate themselves. This process could be accelerated by biotechnological research. It may also be possible to regrow organs through genetic engineering. For example, regenerating brain cells or nerve tissue could offer revolutionary treatments for people with paralysis.

Regeneration of biological organs

Advantages

- **Solving the problem of organ donation:** There are millions of patients waiting for organ transplants in the world, but the number of organs donated is limited. Reproduction of biological organs can solve this problem and reduce the number of patients waiting for organ transplants. Since the regenerated organs will be personalized, the risk of organ rejection can also be significantly reduced.

- **Treatment of diseases and aging:** When organs are regenerated, organ damage due to aging can be treated. Chronic diseases such as heart failure, liver disease, and kidney failure can be treated more effectively. In addition, genetic modifications can prevent or treat diseases in advance.

- **Personalized organs and therapy:** Regenerated organs can be produced specifically for each individual. This can eliminate the possibility of organ rejection in patients waiting for organ donation. Personalized therapy can make medical interventions more effective and safe.

Challenges and ethical issues

- **Technological challenges:** Organ production remains a major challenge, as making organs "fully" functional, locating vascular systems, and building complex biological structures are complex processes. Furthermore, producing human organs in animals can create ethical issues and biosafety risks.

- **Ethical and legal issues:** Biological organ production also raises ethical issues. For example, it is necessary to clarify which organisms will be used to obtain organs, how these organs will be used, and the limits of interventions that will affect human health. At the same time, extending people's lives through organ production and regeneration may increase social inequalities. This may raise questions such as whether organs will be offered only to the wealthy.

• **Social and psychological effects:** The reproduction of biological organs can have profound social and psychological effects on human life. As people approach the idea of immortality, how might these new technologies transform social structure? As people seek to overcome aging or disease, the social impacts of these processes and individuals' relationship with death may change.

Organ production technologies of the future

The reproduction of biological organs has a revolutionary potential for human health. This technology can solve the problem of organ donation, accelerate the treatment of diseases and improve the quality of life of people. However, advances in this field face significant technological, ethical and societal debates and challenges. In the future, a period may begin in which biological organ production will reshape not only medicine but also the entire structure of society.

CRISPR TECHNOLOGY:
GENETIC REVOLUTION

Clustered Regularly Interspaced Short Palindromic Repeats (CRISPR) is considered one of the most important and revolutionary developments in biotechnology. This technology is a tool used to edit and modify genetic material. CRISPR has the potential to revolutionize many areas, from treating genetic diseases to producing more efficient crops in agriculture.

The basis and working principle of CRISPR Technology

CRISPR is a system discovered in bacteria as a natural defense mechanism. When bacteria are attacked by viruses, they can remember the genetic material of the viruses using CRISPR sequences and defend themselves using this information when attacked again. Scientists have adapted this natural mechanism for use in genetic engineering.

Components of CRISPR

1. Cas9 Enzyme: Cas9 is an enzyme that performs the "cutting" function of the CRISPR system. Targeting genetic material, this enzyme can cut the desired point in the DNA chain.

2. Guide RNA (gRNA): Guide RNA targets the specific region to be cut in the DNA. This RNA shows the correct region to the Cas9 enzyme, so the correct genetic material is cut.

Working mechanism

1. CRISPR technology begins with the binding of a guide RNA to a specific region of a target gene.

2. The Cas9 enzyme finds the region directed by the guide RNA and cuts the DNA strand.

3. The cut region is genetically engineered, a new gene can be added, or a faulty gene can be corrected.

4. The cell tries to repair the cut DNA region with its own natural repair mechanisms, where scientists can use various strategies to make genetic changes.

Potential applications of CRISPR technology

1. Treatment of genetic diseases

CRISPR has the potential to revolutionize the treatment of genetic diseases. Genetic diseases such as muscle diseases, cancer, cystic fibrosis, sickle cell disease, and some types of inherited blindness can be treated using CRISPR.

• CRISPR techniques have been applied to correct genetic mutations in the treatment of sickle cell anemia.

2. Cancer treatment

Cancer cells often grow abnormally as a result of genetic mutations. CRISPR offers a potential solution for cancer treatment by targeting the DNA of cancer cells and repairing these mutations.

• It can also be used to modify immune cells to strengthen the immune system.

3. Genetic modification in agriculture

In agriculture, CRISPR can be used to produce productive, disease-resistant or nutritionally valuable products. Genetic engineering can make agricultural products more productive, water-resistant plants can be grown and food security can be increased.

- CRISPR has been used to achieve higher yields in rice, to make tomatoes resistant to diseases or to increase the productivity of wheat.

4. Animal production and regeneration

CRISPR allows the production of genetically modified animals. It may also be possible to produce organs for transplantation in some animal species.

• The production of human organs in pigs is a potential application area for this technology. CRISPR is also being used to cure diseases and increase productivity in animals.

5. Universal vaccine and drug development

CRISPR can be used to analyze the genetic structure of viruses and develop more effective vaccines and treatments by targeting these structures. This technology can be an important tool, especially in the fight against pandemics.

• For example, CRISPR technology is being researched to develop faster vaccine and treatment solutions against COVID-19.

Legal issues with CRISPR technology

While CRISPR technology carries great potential, it also raises many ethical and legal issues:

1. Germline edits (inherited genetic changes)

CRISPR also enables genetic editing of germline cells (egg and sperm cells). This means that genetic changes can be passed on to future generations.

• Such interventions can create ethical issues because permanent changes are made to the genetic structure of individuals. There are also concerns about whether genetic design will create inequality between classes and individuals.

2. Genetic discrimination and social effects

The ability to use CRISPR technology for genetic modification could increase the potential for the wealthy to have healthier, "better" children. This could fuel genetic discrimination and further deepen social inequalities.

3. *Misuse and bioterrorism*

Malicious use of CRISPR could pose serious security risks, such as bioterrorism, particularly the risk of genetically modifying viruses or creating new biological agents that threaten human health.

4. *Bioethics and impacts on human nature*

Altering the genetic makeup of humans could have profound effects on human nature. How such changes would transform individuals' identities and social structures is just one of the ethical questions that must be asked.

CRISPR technology and the future

CRISPR is a technology that has the potential to revolutionize the scientific and medical world. It has a wide range of applications, from treating genetic diseases to producing more efficient crops in agriculture. However, this potential comes with ethical, legal and social issues. Regulations and discussions will continue in the future for the safe, ethical and effective use of CRISPR. Nevertheless, this technology opens the door to a new era for humanity in the field of genetic engineering.

ARTIFICIAL BODY PARTS

Cybernetic limbs and organs are technologies that integrate biological structures and functions with artificial systems. This technology has made significant progress in the field of medicine and is used to artificially regenerate lost or damaged organs and limbs of the human body. Cybernetic organs and limbs are shaped by the combination of developments in biotechnology and robotic engineering. Innovations in this field can provide solutions for disability, organ failure and various diseases.

Cybernetic limbs: Artificial body parts

Cybernetic limbs, commonly known as prosthetics, are artificial organs that replace lost arms, legs, or other organs. However, cybernetic limbs are more than traditional prosthetics; they use artificial intelligence, robotics, and bioengineering to perform biological functions.

Basic features of cybernetic limbs

1. *Movement and functionality*

Cybernetic limbs mimic natural movements by connecting to the user's nervous system. For example, bionic arms can work in a way that mimics muscle movements and can restore subtle sensations such as touch.

• Electromyography (EMG) Technology: EMG analyzes the user's nerve signals and moves the limbs using sensors that detect the electrical activity of muscles.

2. Biotechnological integration

Cybernetic limbs can communicate directly with the brain through integration with nerve endings or muscles. For example, a cybernetic arm can move with the user's thoughts. This includes artificial devices that perceive the brain's signals and perform motor functions.

3. Feedback and sensory technologies

Modern cybernetic limbs can give their users visual, tactile, and even pain feedback. This allows robotic limbs in particular to function in a more natural and biological way. Data from the outside world is transmitted to the brain through sensors.

4. Durability and lightness

Cybernetic limbs are generally made of durable materials (titanium, carbon fiber, etc.), but their lightness is also a priority. This ensures comfort for users during long-term use.

Examples

Bionic arms and hands: Bionic arms and hands can restore the functionality of normal hands by interacting directly with the user's muscle movements. However, some advanced bionic arms can also transmit the sense of touch to the user.

Bionic legs: Bionic legs can enable patients to walk after amputation. These legs are equipped with motors and sensors to mimic the natural stepping motion.

Cybernetic organs
Organ failure and regeneration

Cybernetic organs are artificial organs that can replace organ failure or damage. These organs can perform the functions of biological organs, sometimes with enhanced functions.

Basic properties of cybernetic organs

1. Functional backup: Cybernetic organs could be designed to complement or surpass the functions of biological organs rather than directly mimic them. For example, an artificial heart could be more durable and efficient at pumping blood.

2. Organ connection and integration: Cybernetic organs generally use implant technologies to establish a solid biological connection with the body. Artificial organs must be able to integrate with blood vessels and nerve pathways and be compatible with the biological system.

3. Digital health monitoring: Cybernetic organs can continuously monitor biological parameters. Artificial hearts or kidneys can continuously monitor blood pressure, blood sugar, or other health data and make adjustments as needed.

4. Energy source and power management: Cybernetic organs require a source of energy to function, and this can range from battery-powered devices to systems that generate electricity from body heat.

Examples

• **Artificial heart:** For patients awaiting a heart transplant, artificial hearts can perform the same blood-pumping function as a biological heart, sometimes regulating blood flow and in other cases having the potential to pump more efficiently.

• **Artificial kidney:** For patients who require dialysis treatment, artificial kidneys could mimic the function of kidneys by filtering blood. The development of artificial kidneys has the potential to offer a solution for patients with kidney failure.

• **Artificial eye (bionic eye):** For individuals who have lost their vision, bionic eyes can provide visual information through microchips placed on the retina. This technology can provide partial vision to individuals who have vision loss.

The future of cybernetic limbs and organs

1. Advanced integration and brain-computer interfaces (BCI): In the future, cybernetic limbs and organs will become more advanced and become more sensitive and adaptable by communicating directly with the user's brain waves using brain-computer interfaces (BCI). Thanks to this technology, cybernetic limbs can move with the users' thoughts.

2. Self reapairing systems: Cybernetic organs and limbs, once implanted into the body, could have the capacity to repair and regenerate themselves. Using nanotechnology, these organs could detect and repair damage at the micro level.

3. Aesthetics and biological compatibility: Artificial organs can be designed to be more aesthetically and functionally compatible with biological organs. These organs can function longer than biological organs while maintaining their natural appearance and function.

4. Social problems: The development of cybernetic organs and limbs may also raise ethical and social issues. Modification of the human body may trigger many ethical debates about genetic engineering and biotechnology. Whether these organs will lead to class discrimination or how they will affect human identity are important questions.

SMART SKIN AND TISSUES

Smart skin and tissues are innovative biological materials developed through the combination of biotechnology and nanotechnology. These technologies are designed to mimic or enhance the natural tissue and skin functions of the human body. Smart skin and tissues have the potential to improve not only aesthetics or functionality, but also biological processes. Such innovations could revolutionize medicine, biomedical engineering, and even wearable technology.

Key features of smart skin and tissues

1. *Self repair (regeneration) ability*

Smart skin could have the capacity to repair and heal itself. This feature would allow biological injuries or abrasions to heal quickly. This type of tissue would automatically respond to minor injuries in the body and heal these wounds.

• **Self-healing skin:** Smart skin can rapidly heal injuries caused by injuries. For example, it can be combined with biological systems that detect micro-damage and repair damaged areas. In this way, skin can continuously renew itself.

• **Artificial skin and prosthetics:** Artificial skin can be used in areas such as burn treatment, skin diseases and cosmetic surgery.

Smart skin made with special materials can have sensory properties just like natural skin. This type of skin can respond to physical environments such as temperature, touch and pressure.

2. Biosensor features

Smart skin may have sensors that can collect and analyze biological data. This feature can be used to monitor health status on the body or evaluate environmental factors.

Health monitoring: Smart skin can collect biometric data in real time, such as heart rate, body temperature, sweat rate, blood pressure, etc. This data can provide continuous monitoring to healthcare professionals or send alerts to the user.

3. Environmental sensors

Smart skin could also be sensitive to the external environment. For example, it could detect ultraviolet light and provide warnings to protect the person from sunlight. Sensors could also be placed to detect environmental factors such as toxic gases, chemicals or temperature.

4. Flexibility and adaptability

Smart skin can be physically very flexible, allowing the skin and tissues to move freely and adapt to different parts of the body.

• **High flexibility and durability:** Smart skin can be more durable and elastic than natural skin. This feature allows it to be used in areas that require high flexibility (e.g. joints or surfaces). It can also become more resistant to external factors, for example, resistant to heat, cold or water.

• **Adaptive functions:** Smart skin can automatically adapt to environmental changes. For example, depending on the ambient temperature, the skin can absorb more moisture or respond by sweating.

5. Electronic integration and bioengineering

Smart skin can be integrated with electronic components, so that the skin has both biological and electronic functions. Such technology has great potential for wearable devices or biomedical devices.

• **Wearable technologies:** Smart skin can be integrated into the body like wearable devices. For example, sensors placed on the skin can track body movements or biological data. This could have a range of applications, from monitoring athletes' performance to assessing health status.

• **Bioelectromagnetic fields:** Smart skin could be equipped with devices that can deliver low-level electrical impulses. This could be used for therapeutic purposes. For example, it could provide electrical impulses to relieve muscle pain or help treat neurological diseases.

6. Aesthetic and advanced coloring

Smart skin could have coloration properties and change color according to environmental changes or emotional states. This could be used in both cosmetic and biomedical applications.

• **Color changing skin:** Smart skin can change color depending on environmental conditions (temperature, light, pH value). It can also provide a unique indicator by changing color by sensing emotional states or stress levels. This feature can be used in aesthetics and can also be useful in some medical applications.

7. Potential applications of smart skin and tissues

Medical applications

• **Burn treatment and skin diseases:** Smart skin can be an effective treatment tool for burns and skin diseases (e.g., vitiligo or skin cancer). Thanks to its self-healing and biosensory properties, monitoring and treatment of such diseases becomes more efficient.

• **Prosthesis and artificial skin:** Smart skin can be integrated with prosthetics to produce artificial skin and organs that can function like real skin in place of lost limbs. This allows people using prosthetics to have a more natural experience.

• **Health monitoring and early diagnosis:** Smart skin is a perfect tool for continuous health monitoring. Parameters such as body temperature, blood pressure, heart rate can be immediately transmitted to healthcare professionals. In addition, biological responses to environmental factors can be used to detect early stages of diseases.

8. Military and space applications

Smart skin could be vital in the military or space exploration. Monitoring biological data such as body temperature, humidity, and oxygen levels could increase the functional durability of skin to survive in harsh conditions.

• **Harsh environmental conditions:** Smart skin can be used in military clothing or space suits to monitor the body condition of soldiers or astronauts and adapt accordingly. In addition, additions can be made to protect against heat or radioactive effects from the external environment.

9. Wearable technology and fashion

In the world of wearable technology, smart skin can enable users to monitor their health while also contributing to aesthetics and fashion.

10. Different coloring and style features

Smart skin can be used on clothing or accessories. In the fashion industry, clothes can be made to change color, sense temperature or reflect the mood of the wearer.

11. Environmental impact and energy efficiency

Smart skin could also have energy-efficient features. For example, smart skin surfaces that can generate electricity by collecting solar energy could be used in environmentally friendly wearable technologies.

Technical difficulties

• **Privacy and security:** Since smart skin collects biometric data, the protection of personal information can be a major concern. The security of such technologies will be a significant problem in terms of protecting the privacy of individuals.

• **Side effects and compatibility:** Since smart skin is a combination of biological materials and technology, it may have adverse side effects on the body. Risks such as skin compatibility, skin reactions and allergic responses should be taken into consideration.

- **Social acceptance:** The widespread use of smart skin and tissues may raise ethical questions about biotechnological interventions. Such interventions on the human body may be questionable in terms of social acceptance and individual freedoms.

Smart skin and tissues may represent a major innovation in the field of biotechnology, creating revolutionary changes in health, fashion, military applications and daily life. The potential of these technologies is limitless, both in terms of aesthetics and functionality. However, the challenges of ethics, security and social acceptance must be taken into account in the process of implementing these technologies.

SEEING THE FUTURE THROUGH CONTACT LENSES

Contact lenses are devices made of thin, usually transparent plastic materials that are worn directly on the eye instead of glasses. They are widely used to correct vision defects, but with the advancement of technology in recent years, the functionality of contact lenses has also diversified. These lenses, which are used for both medical and aesthetic purposes, may have much more advanced functionality in the future.

Basic functions of contact lenses

1. *Correcting vision defects*

Contact lenses, used to correct vision defects, treat common eye disorders such as myopia (inability to see far objects clearly), hyperopia (inability to see near objects clearly), and astigmatism (inability to focus properly).

• **Single-focus lenses:** Provide clarity at only one focal length. For example, they allow myopic individuals to see clearly at a distance.

• **Double-focus lenses (bifocal lenses):** These are lenses that provide clarity at both distance and near distances. They are used for those with presbyopia (difficulty in seeing up close with age).

• **Astigmatism lenses:** Correct the incorrect focusing of light due to deformity of the eye.

2. *Aesthetic and colored lenses*

Colored contact lenses, used for aesthetic purposes, can change eye color or create special effects. Groups such as filmmakers, photographers, and cosplay enthusiasts often choose colored lenses.

Advanced technologies and contact lenses of the future

In recent years, the technology of contact lenses has developed rapidly. In addition to the lenses currently used, contact lenses with much more complex functions are expected to enter our lives in the future.

3. Smart contact lenses

Smart contact lenses are an innovation that integrates technology directly with the eye. These lenses contain tiny electronic components and sensors so they can perform functions other than vision.

• **Health monitoring:** Smart lenses can perform health monitoring, such as measuring blood sugar levels, by analyzing the components in eye fluids. This could be very useful for diabetics because it does not require constant blood sugar monitoring. The lenses can provide information about blood sugar levels and alert the user.

• **Monitoring eye fatique:** Smart lenses can detect eye strain or focus issues and send alerts to users, which could be especially useful for those who spend long periods of time looking at screens.

4. Augmented reality (AR) lenses

Contact lenses with integrated augmented reality (AR) technology can project digital information into the wearer's eyes. This allows the wearer to see digital information at the same time as they see the real world.

• **Data display:** AR contact lenses can project navigation information, social media alerts, or other digital data in front of your eyes. For example, you can get directions or see notifications while walking without looking at your phone.

- **Education and health:** The potential of AR lenses in education and medicine is also huge. For example, doctors can see live patient data in front of their eyes, or students can receive interactive training in virtual reality.

5. Lenses for health and treatment purposes

Lenses made using advanced biotechnology not only correct vision defects but can also undertake therapeutic functions.

- **Treatment of mild eye diseases:** Lenses that treat dry eyes, astigmatism, and even some eye infections have been developed, and these treatment methods may become more widespread in the future.

6. Distribution of eye medication

Lenses that carry drugs for eye diseases and deliver them directly into the eye can be used. This could improve intraocular drug treatments in particular.

7. Energy production and storage

Another exciting development is the ability of contact lenses to harvest and store energy from the environment. These lenses can operate on small amounts of energy from the body, eliminating the need to charge the devices.

Ethical and security concerns

As technology advances and contact lenses become more complex, some ethical and safety issues may arise:

- **Data security:** As smart lenses collect health data, data security becomes an important issue. The risk of misuse or leakage of personal health data can pose serious threats to users.

- **Health risks:** Long-term use of smart lenses can cause various health problems in the eye. Wearing lenses for a long time can cause problems such as dryness of the eye, infections or the lens creating effects like a foreign body in the body.

- **Social and ethical issues:** Lenses with advanced AR and data collection capabilities could continuously monitor users' personal information, which could raise ethical questions such as violating societal privacy and individual freedom.

Contact lenses are evolving beyond being a mere vision correction tool to include a range of advanced functions such as health monitoring, augmented reality and biotechnological treatments. In the future, contact lenses with greater functionality, contributing to human health and making daily life easier are expected to become widespread. However, how these technologies will be integrated into society and how ethical issues will be resolved will become clearer as technology progresses.

TELEPORTATION

Teleportation is a concept that is often associated with the idea of instantly transporting an object or person from one place to another, usually found in science fiction works. In the real world, teleportation still seems like a fantasy, but some scientific theories are trying to make the concept more realistic.

Physical teleportation: Quantum-based theories

Teleportation is often associated with quantum physics and quantum mechanics. Two main concepts stand out:

1. Quantum entanglement

Quantum entanglement is a property that allows two or more particles to interact with each other instantly, no matter how far apart they are. This means that changing the state of one particle automatically changes the state of the other.

• **Teleportation and quantum entanglement:** Quantum entanglement could theoretically lay the groundwork for teleportation. Once two particles are entangled, sending one to another location could mean the other "goes" to that new location. But right now, this is just information transfer, not physical transportation of a material object.

2. Transfer of quantum states (quantum teleportation)

Quantum teleportation is the process of transferring quantum information from one object to another. This means that rather than simply transferring the physical existence of an object from one place to another, the object's state and information are transferred from one place to another.

- **Transfer of quantum information:** This process can be accomplished by transferring the state of an atom or a photon from one point to another. However, in this transfer, the matter itself is not physically displaced; only information is transferred.

- **Science fiction and teleportation**

In works of science fiction, teleportation is often depicted as the instantaneous transportation of physical objects or people from one place to another. This is often accomplished with teleportation devices (such as teleportation machines).

- **Star Trek example:** Popular science fiction series like "Star Trek" introduced the idea of teleportation to a wider audience, where the atoms of a person or object are dissolved and rebuilt in another location. This is a scientifically impossible concept, but one that many people dream of.

Difficulties and technological obstacles

The biggest obstacle to teleportation is that the computing power and technological infrastructure required to take the information from objects and transfer it to another location are currently very insufficient. This comes with several significant challenges:

- **Data storage and transport:** The information of each atom must be digitized to reconstruct it exactly. When it comes to teleporting a human, it is necessary to store and process the data of millions of atoms, which is not possible with current technology.

- **Limitations of quantum computers:** The current capacity of quantum computers makes processing big data very difficult. In order to teleport a person, objects or large objects, such a large volume of data would need to be transported quickly and safely.

- **The problem of "destroyin" and "rebuilding" material:** One of the biggest questions about teleportation is how the bodily integrity of an object or a person is preserved. The atomic breakdown and reconstruction of a person or object during the teleportation process is a very complex process and also raises ethical questions. Questions may arise as to whether human identity, consciousness and life can be preserved.

Teleportation and the future

Teleportation, while a possible future goal, remains purely theoretical at the moment. However, as this technology develops, the potential benefits could be extensive:

- **Fast transportation and accsess:** If teleportation were one day possible, distances between people and objects could be reduced to almost zero. Long-distance travel could be accomplished in seconds.

- **Space exploration and humanity:** eleportation could revolutionize space exploration. With teleportation technology, humans could travel between planets in a very short time. This could be a great discovery for humanity and open up new areas of life.

- **Logistics and supply handling:** Teleportation could also transform the logistics sector. Transporting large and heavy materials would no longer require physical transportation and would become much faster and more efficient.

- **Global interaction and connection:** Teleportation makes global interactions much easier. People can come together instantly and interact more closely with cultures in different places.

Could teleportation be real?

Teleportation is currently a purely theoretical concept, in the realm of science fiction. However, the rapid development of quantum physics and quantum computers may one day allow us to take the question of whether teleportation is possible more seriously. While there are currently technological and ethical hurdles to teleporting people or objects, future scientific developments may make significant strides in this regard. Teleportation may one day become a reality with new discoveries in science, but only time will tell.

AI IN THE FUTURE, ROBOTS, AND AUTONOMOUS SYSTEMS

HUMANOID ROBOTS

The integration of robots into human life is a rapidly progressing process. Today, robots have begun to be used in industrial production, health, security, the service sector and social areas. However, in the future, humanoid robots may become an inseparable part of daily life.

Potential roles of robot humans in our lives

1. Helper robots in homes

• They can undertake tasks such as cleaning, cooking, shopping and childcare.

• Social robots that can provide emotional support can be companions for elderly individuals living alone.

2. Health sector

• Nurse robots in patient care.

• High-precision robots that assist surgeons in surgeries.

• Assistive robots in physical therapy and rehabilitation processes.

3. Education

• Robot teachers who provide individual education to students.

• Support for teaching with virtual and augmented reality.

4. Service and retail sector

• Robots that serve as service personnel in restaurants and cafes.

• AI-powered guide robots that provide customer service in stores.

5. Security

• Robots that perform security patrols.

• Robots that provide public safety in cities and detect potential threats.

6. Industry and logistics

• Complete automation of factory production processes.

• Robot systems that optimize storage and shipping processes.

7. Personal relationship

• Robot partners who can communicate and form emotional bonds with humans.

The effects of robot people on life

Positive effects

• Relieving workload and saving time.
• Making life easier for the elderly and disabled.
• Increased efficiency in production and service sectors.
• Individualized learning opportunities in education.

Negative effects

• **Unemployment:** Risk of losing their jobs, especially for people doing routine work.

• **Security problems:** Faulty robot decisions or malicious use.

• **Ethical issues:** Human-robot relations and identity problems.

• **Dependence:** Human social isolation and over-dependence on robots.

Integration of robot humans into life

1. Near future (10-20 years)

• Social and service robots will become widespread.

• Robot use is expected to increase in homes and healthcare services.

2. Medium-term future (30-50 years)

• Robots will look more like humans and will be able to engage in more complex emotional interactions.

• Robot-human co-living spaces may emerge.

3. Long-term future (50 years and above)

• Robots integrated with biological organs (cybernetic humans) may come to the fore.

• The difference between humans and robots may become blurred.

Result: Intertwined life

Robots will increasingly become part of our lives, especially in the areas of work, health, security and social life. However, if the ethical, social and legal dimensions of this integration are not carefully managed, it is possible for both individuals and societies to encounter serious problems. The relationship between humans and robots seems to be one of the dynamics that will radically change the structure of society in the future.

UNMANNED VEHICLES (AUTONOMOUS VEHICLES)

Unmanned vehicles (UAs) are vehicles that can perform certain tasks without direct human intervention. These vehicles use artificial intelligence, sensors, robotic technologies, and autonomous software to perceive their surroundings and act on their own. Unmanned vehicles span a wide range of industries, from automobiles to aircraft to marine vessels, and are expected to revolutionize many industries in the future.

Types of unmanned vehicles

Unmanned land vehicles (autonomous vehicles)

Autonomous vehicles are ground vehicles that can travel on their own without the need for a driver. These vehicles use sensors, cameras, LIDAR (light detection and ranging), radar, and artificial intelligence to perceive the environment and make decisions.

• **Automobiles:** Tesla, Waymo and other tech companies are developing and testing driverless vehicles that navigate safely by analyzing traffic, road conditions and obstacles.

• **Logistics and transportation vehicles:** Autonomous trucks could revolutionize the logistics and transportation industry, reducing the need for human drivers and increasing the efficiency of long-distance transportation.

• **Shipping and delivery vehicles:** Especially with the growth of e-commerce, autonomous vehicles can be used to make small deliveries. For example, autonomous vehicles delivering cargo in a city can deliver packages quickly and safely.

Unmanned Aerial Vehicles (UAV / Drones)

Drones are unmanned aerial vehicles that are typically guided by an operator during flight, although more advanced models can also operate autonomously.

• **Commercial use:** Drones are used in a variety of sectors: cargo transportation, plant health monitoring in agriculture, inventory management, mapping and reconnaissance, and are rapidly spreading in areas such as.

• **Military and defence applications:** In the military field, drones are used in a wide variety of tasks, including reconnaissance, target detection, bombardment and material transportation.

• **Emergencies and search and rescue:** Drones can also be used in lifesaving missions by quickly collecting data or delivering relief supplies in disaster areas.

Unmanned marine vehicles (Autonomous ships and boats)

Autonomous marine vehicles are used in sea transportation, research and exploration missions. These vehicles can follow the route at sea, perform various tasks and travel safely.

• **Autonomous merchant ships:** Autonomous shipping could revolutionize the shipping industry, eliminating the need for human crews and improving safety and efficiency.

• **Discovery and research:** Autonomous marine vehicles used for deep-sea exploration and scientific research can be used in tasks such as underwater mapping, ocean research and oil/gas drilling.

Unmanned space vehicles

Unmanned spacecraft are vehicles used for space exploration that can operate without the need for humans. These vehicles are used to explore distant planets, mine asteroids, search for life in space, and other scientific missions.

- **Mars and moon exploration vehicles:** NASA and other space agencies are sending autonomous rovers to Mars, which can explore the planet's surface and provide essential information for human exploration.

- **Research in space:** Research into deep space objects such as asteroids or comets can be carried out by autonomous space vehicles.

Advantages of unmanned vehicles

- **Safety:** Unmanned vehicles can improve safety by eliminating human error. For example, driverless cars can reduce traffic accidents because human error accounts for most accidents. Additionally, autonomous drones or marine vehicles can operate in dangerous areas, such as war zones or disaster areas.

- **Efficiency and time saving:** Autonomous vehicles can operate 24/7 and perform tasks continuously without the need for humans to rest. This provides a huge efficiency increase, especially in areas such as transportation and logistics.

- **Cost savings:** Unmanned vehicles can reduce costs by reducing the need for manpower. In addition, their efficient and fast operation can optimize the use of time and resources.

- **Use in hazardous tasks:** Dangerous work can put people at risk. Autonomous vehicles could protect lives in mines, on battlefields or at dangerous chemical spill sites.

Challenges of unmanned vehicles and ethical issues

- **Safety and margin of error:** Unmanned vehicles can be made completely safe with technological developments, but any technical failure or system error can lead to serious accidents. In addition, the sensors and algorithms developed for autonomous vehicles to perceive the environment correctly must work perfectly under all conditions.

- **Job loss:** The use of unmanned vehicles could replace the workforce in many sectors. This could lead to job losses and create social problems, especially in areas such as transportation, logistics and defense.

- **Ethical and legal issues:** The decision-making processes of autonomous vehicles may raise ethical issues. For example, what decision should an autonomous vehicle make in the event of an accident? Can autonomous vehicles make mistakes that human-powered vehicles do not? Such questions have raised ethical issues regarding the design and use of vehicles.

- **Syber security risk:** Because unmanned vehicles are connected to the internet, they can be vulnerable to cyberattacks. Malicious actors can hack into the vehicles and take control, posing serious security risks.

Unmanned vehicles offer revolutionary technology that promises major changes in the fields of transportation, logistics, reconnaissance and defense. With the development of autonomous vehicles, advantages such as safety, efficiency and cost savings can be provided, while difficulties such as ethical, security and labor issues can be encountered. As technology advances, these vehicles will continue to become more widespread, but solving social, legal and ethical problems in this process will be of great importance.

THE COMBINATION OF AI AND EMOTIONAL INTELLIGENCE

Cybernetic limbs and organs are some of the most exciting examples of technology merging with biology. These technologies can restore lost functions, improve the quality of life of individuals suffering from organ failure, and in the long term, make it possible to overcome biological limits. However, the ethical, security, and societal implications that come with these advances must also be carefully considered. In the future, cybernetic technologies may transcend the natural limits of the human body, providing a more efficient, durable, and flexible life.

Expanding emotional capabilities: New boundaries between humans and technology

Expanding emotional capabilities could allow people to more deeply understand and manage the emotions they feel and experience, and to communicate more empathically with others. This concept offers new ways to enhance or alter emotional experiences, thanks to advances in biotechnology, artificial intelligence, and neuroscience. Various tools and technologies are being developed to help people both enhance their own emotional worlds and establish healthier and more productive relationships with others.

Expanding emotional abilities

1. Emotional Intelligence (EQ – Emotional Quotient)

By combining AI with emotional intelligence (EQ), algorithms with human-like emotional capacity can be created, providing more sophisticated technology to understand and respond to human emotions.

- **Emotional recognition systems:** AI can analyze factors such as tone of voice, facial expressions, body language, and biometric data to determine emotional states. This allows for an accurate assessment of users' mood.

- **Empathy and emotional communication:** AI-based systems can be designed to understand users' emotional states and develop empathic responses. This could be particularly useful for therapeutic applications, customer service, or social robots.

2. Brain-computer interfaces (BCI) and emotional interactions

Brain-computer interfaces (BCI) are systems that read and interpret the brain's electrical activity. These technologies can be used to monitor and modify people's emotional states.

Monitoring emotional states: BCIs can monitor the brain's emotional responses and provide deeper insights into the user. For example, a BCI can detect when a person is feeling stressed or depressed.

Emotional feedback and improvement: By providing emotional feedback, BCIs can help people stabilize their mood or promote a positive mood, allowing for conscious interventions to manage emotional states.

3. Neuroenhancement and emotional expansion

Neuroenhancement is an intervention that aims to improve or change brain functions. This has great potential to improve emotional abilities.

- **Brain stimulation and emotional healing:** Technologies such as Transcranial Magnetic Stimulation (TMS) can be used to treat depression, anxiety or other emotional disorders by targeting specific areas of the brain. Such neuroinjection techniques can play an important role in improving emotional balance.

- **Positive emotional reinforcement:** Neuroenhancement techniques can help the brain experience positive emotional states, such as happiness, compassion, or empathy, more intensely. This allows people to better manage their moods and increase their overall emotional intelligence.

4. Genetic regulation and emotional abilities

In the future, with tools like genetic engineering and CRISPR, emotional abilities could also be altered or enhanced. This has the potential to change the emotional nature of humans.

- **Emotional genetic modification:** Genetic engineering can alter people's brain chemistry and neurotransmitter levels, for example by making genetic modifications that make them more empathetic, increase the release of the happy hormone, or improve their ability to cope with stress.

- **Improved emotional regulation:** We can increase emotional balance through genetic tweaking so that people can better manage conditions such as extreme stress, anxiety or depression. Such changes can lead to a healthier, more balanced lifestyle.

5. Biotechnology and emotional communication tools

Biotechnological devices can be used to increase emotional intelligence and empathy. Wearable devices can monitor people's moods and use this for personal development or health purposes.

- **Wearable technologies that enhance emotional feeling:**

Devices such as smart watches or EEG caps can monitor people's moods and provide information about emotional states. These devices can provide suggestions to improve the user's emotional balance or provide guidance for coping with stressful situations.

- **Emotional intelligence training applications:** Biotechnological devices can be tools to support emotional intelligence training, especially for children and adolescents. These devices can help children develop empathy, compassion or emotional regulation skills.

Potential applications of expanding emotional capabilities

- *Therapeutic uses*

Expanding emotional capabilities could revolutionize psychological health in particular. Therapeutic robots, AI-powered counseling services, or emotional monitoring devices could be effective in treating depression, anxiety, post-traumatic stress disorder (PTSD), and other mental disorders.

- **Emotional support robots:** Robots and AI can be used as personal therapists or emotional support providers, facilitating therapeutic processes by providing emotional support, especially for people experiencing loneliness or depression.

- *Social relations and empathy*

Expanding emotional abilities can help create healthier and more understanding social relationships by increasing empathy between individuals. This can lead to greater understanding and tolerance, especially in society.

- **Emotional intelligence training programs:** Emotional intelligence training in schools and workplaces can help individuals better understand themselves and others. Developing empathy can reduce conflict between people and create a healthier society.

Advanced human-machine interactions

Emotional expansion could make human-machine interactions more natural and productive. AI systems that can respond emotionally like humans could enrich the experience of humans interacting with robots.

- **Emotional robots and interactions:** Robots that can deliver human-like emotional responses could be used in education, customer service, and even outreach, allowing people to form deeper, more meaningful relationships with machines.

Challenges and problems

1. Emotional manipulation

Expanding emotional capabilities could potentially increase the risk of emotional manipulation. The ability to change people's emotional states through artificial intelligence and biotechnology could lead to ethical issues, such as the potential for commercial exploitation of individuals' moods or the misuse of personal information.

2. Identity and humanity issues

Altering emotional abilities can raise philosophical questions about human identity and nature. Artificially altering the emotional nature of humans can open up debates about free will and the perception of humanity.

The future of expanding emotional capabilities

Expanding emotional capabilities can improve people's quality of life, promote empathy, and have positive effects on individual psychological health. However, it is important to carefully consider the ethical and societal implications of these technologies. With the combination of artificial intelligence, neuroscience, and biotechnology, it may be possible to enhance people's emotional intelligence and create a healthier society.

THE COMBINATION OF HUMAN AND ANIMAL ABILITIES

Biological synthesis refers to the integration of human and animal capabilities through the combination of fields such as genetic engineering, biotechnology, and artificial intelligence. Such synthesis could allow humans to acquire certain biological characteristics and capabilities in animals, or to surpass the biological limitations of humans and mimic animal characteristics. There are many different capabilities and biological characteristics between humans and animals, and synthesizing these capabilities could potentially open new doors for both medical and human developments.

Integrating animal abilities into humans

1. *Enhanced sensory sensitivity*

Animals have sensory abilities that may be inaccessible to humans. For example, some animals can see ultraviolet light, while others can hear sounds at much wider frequency ranges. Bringing these abilities to humans is one of the fundamental areas of biological synthesis.

• **Vision and hearing:** For example, some seabirds and insects can see ultraviolet light, while humans can only perceive the visible range. Genetic engineering could give humans these sensory abilities. In addition, bats' superior hearing (echolocation) abilities could become possible for humans.

• **Sense of smell:** Animals can use their sense of smell in a much more advanced way than humans. For example, dogs have a much more sensitive sense of smell than humans. If biotechnology were to enhance this ability in humans, new applications could emerge; for example, diagnosing diseases by smell.

2. Speed and power

Some animals have extraordinary physical abilities; for example, cheetahs can run very fast and kangaroos can leap great distances. Biologically enhancing such physical abilities for humans could significantly increase their physical capacities.

- **Muscle strength and endurance:** Humans have muscular structures that work within certain limits, but some animals stand out with their incredible speed or strength. Such characteristics can be integrated into humans through genetic engineering. For example, genetic modifications can give humans powerful muscular structures, such as lions.

- **Running speed:** Reaching the speed of fast running animals like the cheetah may be possible through biological interventions on muscle structure and metabolism. Increasing the speed capacity of humans may provide benefits in terms of health as well as physical performance.

3. Ability to regenerate

Some animals have the capacity to regenerate lost organs or limbs. This ability allows the creation of a biological synthesis that surpasses the biological limits of humans and allows the regeneration of damaged organs and tissues.

- **Biological regeneration:** Animals such as planarians and sea stars can regenerate lost body parts. Humans may gain this ability through biological engineering. For example, allowing the human body to naturally regenerate damaged organs, nerve cells or limbs could be one of the biggest revolutions in healthcare in the future.

- **Organ regeneration:** Just as deer renew their antlers every year, humans can regenerate their organs or tissues. By improving such regeneration processes through genetic engineering, the need for organ transplants may be reduced.

4. High temperature and cold resistance

Some animals have extraordinary abilities to survive in extreme conditions. If humans acquired such biological abilities, they could increase their capacity to survive in different climatic conditions.

• **Cold resistance:** Polar bears or penguins are animals that can survive in cold environments. Biotechnological interventions can be made to make humans more resistant to cold weather conditions. This can make it easier to survive, especially in polar regions or during space travel.

• **Heat resistance:** Desert animals are organisms that can withstand high temperatures. The genetic structure of humans can be altered to make them more resistant to heat. This way, humans can be more resistant to drought and extreme temperatures in hotter climates.

5. Dual intelligence and bioelectric systems

Animals can sometimes exhibit extraordinary cognitive abilities. For example, some dolphins, monkeys, and some species of birds have highly developed social intelligence and problem-solving abilities. Such cognitive abilities can be acquired by humans or combined with human intelligence.

• **Social intelligence among animals:** Animals such as primates and dolphins can form complex social structures and cooperate. Humans can develop such social intelligence abilities. For example, we can strengthen empathy, cooperation, or decision-making processes among humans by drawing inspiration from the social intelligence of animals.

• **Bioelectric signals:** Animals can effectively use electrical signals in their nervous systems. Human brain activities can be more integrated with bioelectric systems, thus enabling faster and more efficient brain-machine interaction.

Ethical issues

While biological synthesis enables the integration of human and animal capabilities, it can also raise ethical issues. Modifications to human genetics require deep thought about identity, free will, and the impact on society beyond biological boundaries.

• **Ethical aspect of genetic manipulation:** Biological manipulations on humans, such as creating genetic differences or imparting certain abilities, may raise ethical issues. Such changes may have negative effects in terms of human rights and social equality.

• **Animal rights and human-artificial life relations:** Giving humans animal abilities may raise the need to draw boundaries between animal rights and human rights. Furthermore, the differences between the "natural" state of humans and the state altered by artificial interventions, and the social acceptance of these differences, may be an important topic of discussion. Biological synthesis has great potential for the integration of human and animal abilities. Such biotechnological advances can transform human biology, improve health, and improve living conditions. However, the ethical questions brought by such developments may also bring social and philosophical discussions. Biological synthesis is not only a scientific innovation, but also a field that will allow humans to redefine themselves and transcend biological boundaries.

THE POTENTIAL ROLE OF AI IN THE LAW AND JUSTICE SYSTEM

a. Data analysis for legal research and litigation

Artificial intelligence can be used as a tool that can perform rapid and in-depth analysis of data and documents in legal cases. Lawyers and judges can review similar cases, precedents, laws and legal documents more quickly through AI-powered systems.

- **Legal research and case law litigation:** By analyzing large data sets, AI can quickly review past cases, similar legal arguments, and previous court decisions. This can help lawyers and judges make more informed and faster decisions.

- **Probability calculations and prediction models**

By learning from past cases and court decisions, AI can predict the likely outcome of a case, allowing lawyers and clients to better tailor their litigation strategies.

- **Prediction models:** AI algorithms can make decision predictions by considering the impact of certain factors in a case, such as predicting the punishment for certain crimes in a criminal case or calculating the potential compensation amount in a commercial case.

Smart contracts and automated agreements

Blockchain and AI integration could allow for the automation of agreements and contracts. Such systems would automatically enforce a contract when certain conditions are met between the parties.

- **Smart contracts:** Smart contracts will automatically enter into force when certain conditions are met, eliminating the need for any intermediary between the parties. This can be particularly effective in commercial, property and rental agreements.

Legal advice and accessibility

Artificial intelligence can also play an important role in the field of legal consulting. In particular, it can be used to provide low-cost, fast and efficient consulting services.

- **Chatbots and legal asistance:** AI-powered chatbots can help individuals quickly obtain legal advice. In particular, digital consultants can come into play for basic legal issues and applications.

Automations of court rulings and trial processes

Some simple cases can be automated by AI. This can reduce the burden on the courts and provide a faster resolution process. AI can act as a judge, especially in matters such as commercial law, traffic tickets, and small claims.

- **Automated judgement and decision making:** AI can make decisions based on certain legal parameters, but there will still be areas where human judgement will be needed to ensure ethics and fairness.

Judicial independence and human factor

AI can speed up the decision-making process by providing objective data and logical arguments, but the human factor, empathy, values and ethical decisions will still play an important role. Judges need to make decisions that go beyond the law and that are based on the values of society and the delivery of justice.

- **The role of human judges:** The full decision-making capability of AI could pose a major ethical challenge. Human judges are capable of making emotional and ethical decisions in a social context, preserving the human aspect of law.

Ethical and legal issues of artificial intelligence and the judicial system

1. Accessibility of justice

AI may become a resource that only high-net-worth individuals and large law firms can access. This could lead to an unequal distribution of legal services. When digitalization does not provide the same opportunities for everyone, access to justice could be at risk.

2. Algorithmic biases

Because AI is trained on historical data, it can learn biases in data sets and reflect them in its decision-making processes. This can lead to unfair outcomes based on race, gender, or socioeconomic status.

- **Prejudice and justice:** Historical biases in the data that AI systems are trained on can also influence future decisions. For example, past crime rates or sentencing practices can have the wrong impact on future decisions.

3. Human rights and privacy

Artificial intelligence can analyze the behavior of individuals using big data. However, such data use can violate personal privacy rights. There is also the risk that the data will be misused or misused.

- **Data security:** In legal processes, the privacy and rights of individuals must be protected. It is important to provide security measures during the storage and analysis of digital data.

4. AI and judicial indepence

The integration of AI into the legal system could threaten judicial independence. Algorithms and data sets could be manipulated by governments or large corporations, making it harder for judicial decisions to be fair and independent.

5. Legal liability

When an AI makes any ethical or legal violations while making decisions, the question of who will be held accountable arises. Which party will be held responsible for a wrong decision made by an AI can be a complex issue.

Future judiciary, law and justice system: Artificial intelligence and digital transformation

It is anticipated that the future judiciary, law and justice system will be shaped by the digitalization and integration of artificial intelligence. Artificial intelligence can increase efficiency in the field of law, speed up trials, reduce errors and be an important tool in the provision of justice. However, there will also be challenges such as ethics, security and human rights. Artificial intelligence can transform the future law and justice system. However, there will be many ethical, security and social control issues that need to be considered during this transformation. The integration of artificial intelligence into legal processes has the potential to create not only a faster and more efficient system, but also a fairer and accessible legal system. However, before an artificial intelligence powerful enough to replace the human factor in the provision of justice can be achieved, these systems will need to be designed and supervised in a reliable, transparent and equitable manner. Law is not only about the application of rules, but also the protection of human rights, values and justice. In this context, using artificial intelligence as a supporting tool, but keeping the final decisions in the hands of humans, will preserve the spirit of the law.

- Will humans and robots be able to marry in the future?

2. Scince fiction themed questions

- Will there be mind-reading AIs?
- Will AI take over the world and form governments?
- Will people start worshipping AIs?

3. Absurd and humorous questions

- Can AI tell me lottery numbers?
- Will cats use AI in the future?
- Can AI brew coffee and chat?

4. Inconsistent questions in itself

- Can AI change the past in the future?
- Can AI interfere with my dreams?
- Will AI exist forever?

5.Questions about conspiracy theories

- Can AI communicate with aliens?
- Could AI turn humans into zombies in the future?
- Are governments conducting secret experiments with AI?

These types of questions are often asked under the influence of science fiction movies or for humorous purposes. Although some may seem funny, it is important to ask more logical questions to try to understand the limits and potential of artificial intelligence.

IV

FUTURE ECONOMY AND WORKING LIFE

DIGITAL MONEY

Digital money is a currency that exists only in a digital environment without a physical entity. Digital money is a structure that has gained an important place in the financial world with rapidly developing technology and aims to replace traditional monetary systems. This form of money includes different types of blockchain technology, cryptocurrencies and digital currencies.

Types of digital currencies

1. *Cryptocurrences*

Cryptocurrencies are decentralized and encrypted digital currencies. They are traded on digital platforms outside the control of traditional banks. The most well-known example of cryptocurrencies is Bitcoin, but there are other types such as Ethereum, Litecoin, Ripple.

• **Bitcoin (BTC):** Bitcoin is the first cryptocurrency created by Satoshi Nakamoto in 2009. It is called digital gold due to its decentralized structure and limited supply.

• **Ethereum (ETH):** Ethereum is not just a digital currency, but also a platform where you can develop smart contracts and decentralized applications (dApps). Ethereum's cryptocurrency, ETH, is used to perform transactions on the blockchain.

• **Other cryptocurrencies:** Other cryptocurrencies such as Litecoin, Ripple (XRP), Binance Coin (BNB) are also digital currencies with different features. Each has different uses and advantages.

- *Centralized digital currencies (CBDC)*

Centralized Digital Currency (CBDC) are digital currencies issued by governments and central banks. Unlike cryptocurrencies, they are managed by a central authority.

- **China's digital yuan (e-CNY):** China, as a pioneer in the field of digital currency, has developed the digital yuan (e-CNY), a central bank-issued currency that is fully controlled by the state.

- **The European Union and other countries:** The European Central Bank and many countries are working on CBDCs, which could accelerate digital payment systems and increase financial inclusion.

2. Crypto assets and tokens

Crypto assets are units of value in digital systems, and besides cryptocurrencies, various tokens also fall into this category.

- **Tokens and ICOs:** Tokens are digital assets created on the blockchain. These tokens can provide transactions on a certain platform or access to a certain service. ICOs (Initial Coin Offerings) are investment campaigns for new tokens.

Advantages of digital currencies

1. Fast and cheap transactions

Digital currencies, especially cryptocurrencies, allow money transfers to be made faster and with lower transaction fees compared to traditional banking systems. This offers great convenience in cross-border money transfers.

2. Decentralization and freedom

Cryptocurrencies are not dependent on a central authority. This allows users to have complete control over their assets. It also facilitates access to financial services for individuals who do not have access to banking systems.

3. Security and transparency

Blockchain technology provides security by encrypting digital currency transactions. Additionally, transparency is increased because all transactions are recorded in a public ledger.

4. Financial participation

Digital currencies provide financial inclusion to millions of people around the world who lack access to financial services. Especially in developing countries, digital currency solutions can expand financial access.

5. Protection agains inflation

Some digital currencies provide protection against inflation because their supply is limited. For example, Bitcoin has a limited supply, which can help it gain value over time.

Challenges and risks of digital currencies

Volatility

The value of cryptocurrencies is often highly volatile. For example, cryptocurrencies such as Bitcoin or Ethereum can often experience sudden changes in value. This can make it difficult to use digital currencies as a means of daily trading and payment.

Regulations and legal uncertainties

Digital currencies still operate in regulatory uncertainty in many countries. Governments have yet to clarify how they will regulate cryptocurrencies and implement tax policies. There may also be bans or restrictions on the use of digital currencies.

Security issues

Digital currency wallets, exchanges and platforms can be vulnerable to cyber attacks. Therefore, the security of digital currencies is a major concern. The risk of hacking can lead to users losing their digital assets.

Abuse and money laundering

Cryptocurrencies can also be used for illegal activities due to their anonymous features. Activities such as money laundering and illegal trade can be carried out with transactions made with digital money.

Technological dependency

Digital currencies are built entirely on digital infrastructures, which increases the dependence on technology. Any technical failure or system crash can lead to major financial losses.

The future of digital currencies

Digital currencies are likely to play a major role in the future financial system. While cryptocurrencies will allow the financial system to become more decentralized, CBDCs will allow governments to control digital currencies. The increased prevalence of digital currencies could speed up financial transactions, reduce costs, and make financial services more accessible to a wider audience.

However, the future of digital currencies will be shaped by regulatory frameworks, security measures, and social acceptance. In order for digital currency systems to be more widely adopted, legal uncertainties will need to be resolved, security will need to be increased, and technological infrastructure will need to be strengthened.

E-COMMERCE, NORMAL TRADE, INDUSTRIALIZATION AND TRADESMANSHIP

Technological developments and changing social dynamics will radically transform the future forms of trade. The growth of e-commerce, new approaches to industrialization, and the evolution of craftsmanship and physical trade will affect each other in the future. Here are the possible directions of these changes:

The evolution and future of e-commerce

• Fast and personalized shopping experience: E-commerce will become even more personalized thanks to artificial intelligence and machine learning. Users will receive product recommendations according to their personal preferences while shopping, and their shopping experience will be completely personalized.

• Virtual stores and holograms: Virtual stores and holographic showcases, which are the digital equivalents of physical stores, will emerge. When shopping online, customers will be able to see products in art galleries or holographic showcases as if they were in a physical store.

• Voice and visual shopping: Users will be able to shop with voice commands and search for products through voice assistants (Alexa, Siri, etc.). In addition, video-based shopping will become more common, and customers will watch products online and make instant purchases.

• Drone and autonomous delivery: Delivery processes will be accelerated with drones and autonomous vehicles. This will provide a fast and efficient shopping experience, such as same-day or short-term delivery of products.

The future of physical trade and tradesmanship

• Digitized tradesmanship: Tradesmen will be able to expand their local businesses using digital tools. Small businesses will be able to appeal to a wider audience through social media platforms and online marketplaces. At the same time, AI-powered digital marketing tools will help tradesmen reach their customers more effectively.

• Social commerce: Shopping made through social media and digital platforms will become more common. Tradesmen will increase their sales with live broadcasts, social interactions and customer feedback on social media.

• Transformation of physical stores: Physical stores will now go beyond just selling products. These stores will transform into experience centers; customers will come together physically, experience and test products and then complete their shopping online.

• Inventory management with artificial intelligence: Tradesmen will be able to manage their inventory in real time using artificial intelligence. They will also be able to quickly update their stocks in line with customer demands and trends.

The future of industrialization and manifacturing

• **Digital production and 3D printing:** 3D printing technology will create a major change in industrialization. Now, many productions will be done with on-site production rather than traditional production lines. This will both reduce production costs and offer faster and more flexible solutions with local production.

• **Automation and robots:** With Industry 4.0, robots and

artificial intelligence-based systems will make production processes faster and more efficient. The need for manpower will decrease, but the development of these robot technologies will create new business areas.

• **Modular production:** Using modular production systems, industries will be able to produce each product in a way that is customized according to personal demands. This will make it possible to create special products according to personal preferences and make industrialization more flexible.

• **Artificial intelligence-supported design and optimization:** Production processes and designs will be optimized with artificial intelligence. While designs will become more efficient, the workforce will focus on more creative processes.

New trade models and the global economy

• **Blockchain and digital currencies:** Blockchain technology will increase the security of trade. Cryptocurrencies and digital payment systems can become the basic payment instrument for all trade. This will enable international trade to be conducted faster, more securely and more transparently.

• **Strengthening local economies:** The increase in e-commerce will lead to the strengthening of local production and local consumption processes. People will turn to more local products and small businesses will be able to open up to global markets with digital tools.

• **Shared economy and subscription models:** Shared economy (e.g. car sharing, home sharing) and subscription-based trade models will become more prominent. Instead of renting products, people will rent them as needed and move towards a more sustainable economic model.

• **Green trade:** Sustainability and environmental awareness will increase. Companies will produce eco-friendly products and social responsibility projects will become more widespread. Green energy and waste management will become one of the basic elements of trade.

The result: changing trade dynamics

In the future, the boundaries between e-commerce, physical commerce, industrialization and tradesmen will become increasingly blurred. Digitalization will increase the speed of trade while also offering a more personalized, flexible and participatory shopping experience. Technology will make tradesmen more efficient and trigger more sustainable and local production models in the industrialization process. All of these will lead to the reshaping of local trade together with the globalizing world and societies will evolve towards a more digital, sharing and socially responsible understanding of trade.

IMPORTANT PROFESSIONS OF THE FUTURE

1. AI engineering

Artificial intelligence is used in every area of our lives. Artificial intelligence, which is constantly developing and will continue to develop, offers many software that offer functional solutions to our daily lives. When we consider that we have high expectations from artificial intelligence, we can immediately understand that there is still a lot of work to be done in this area and a lot of software to be developed. Area of use Artificial intelligence engineering, which will spread to dozens of different sectors, is definitely one of the professions of the future.

2. Robotics engineering

Robots from science fiction series seem to be slowly entering our real lives. In this technology created by using artificial intelligence technology, robots can learn functions such as seeing, touching and perceiving. There will definitely be a need for a new workforce for these tools, which will undoubtedly be used more in the future.

3. Drone piloting

One of the new professions that has emerged with developing technology is drone piloting. The use of unmanned aerial vehicles in many different sectors in recent years has increased interest in this profession. We can say that as the number of these devices increases, more pilots will be needed. Unmanned aerial vehicles are very close to the applications that we will frequently use and need in our daily lives. The drone delivery model, which Amazon, one of the world's largest online purchasing companies, first introduced in 2016 but has not yet fully put into use, shows us that this technology will become a part of our lives in the very near future.

4. Synthetic biology expertise

The most important food items consumed in the world are red meat, white meat and fish. If we consider that the world population is rapidly increasing and these resources are limited, we can see how big a problem humanity will face in terms of food. Being able to produce meat in laboratory conditions has the potential to solve many of the world's problems in this regard. In addition to preventing a possible food crisis, this situation can also significantly reduce the amount of carbon emissions, which is another of the world's biggest problems.

5. E-Sports coaching

Online games have become quite popular in recent years and major tournaments are being organized in this field. Gaming is no longer a hobby and there are even federations in this field. The heated competition in this field has also increased the need for coaches who undertake important tasks such as motivating e-sports players and helping them develop strategies. It can be one of the best professions with a future, especially for individuals interested in the world of e-sports.

6. Human resources expertise

Human resources expertise is currently considered one of the best professions today, and we can say that this position will continue to be important in the future. This area, which is especially important for large companies, will continue to be important in the future. Many other elements that constitute company discipline, such as payroll, recruitment, termination of employment contracts and organization, are under the responsibility of the human resources expert.

7. Software development

One of the most demanded professions of recent years, software development is a profession that has a place in many different sectors. Software developers can work for large companies or work as freelancers and earn satisfactory wages. The most important thing a software developer needs is coding skills; It can be learned through school or courses or on your own. The most important thing here is that you can do your job well. We can easily say that it is the profession of both today and the future.

8. Blockchain expertise and development

Blockchain experts and developers design blockchain protocols and undertake the task of creating smart contracts. Although it is normal to think of formations such as Bitcoin when it comes to blockchain, it is necessary to know that it is not limited to this. This technology aims to make the functioning of the digital world more secure and efficient. Blockchain experts and developers are currently widely present in many sectors, especially blockchain and banking. We can say that this profession will be one of the most important professions of the future in Turkey and the world.

9. Game development

A game developer is a software developer who creates games of different types and concepts and brings them together with users. A game developer is present at every stage of the game's production process and continues to carry out the development process even after the game reaches the user. They have duties and responsibilities such as ensuring the technical development of the game and fixing software errors, determining new game content and gamifying the determined content, and following and adapting to technology and conducting research and development. The increase in the time spent on the internet and the fact that online games have become a growing sector show that this profession will be a sought-after and needed business in the future.

10. Digital marketing expertise

Digital marketing expertise, which is among the most popular professions of the future, has become one of the most preferred professions in our country. As companies become increasingly digitalized within the developing technology and increase their focus in this direction, this paves the way for a large employment creation in this field.

11. Data science

A data scientist is someone who is part computer scientist, part mathematician, and can follow and spot trends to some extent. It should also be said that data science is currently a very popular and highly paid profession. Data scientists can also be defined as technically skilled, new generation analytical data experts who solve complex problems and discover which problems need to be solved. It is no secret that this emerging profession will be more popular in the coming years.

12. Cybersecurity expertise

Cybersecurity experts, who are currently in great need in cybersecurity companies, can make serious profits. Many large companies also establish their own cybersecurity teams and employ cybersecurity experts in these teams. The importance of cybersecurity increases day by day as developing technology, especially in the field of informatics, and creates new opportunities in this field. Being equipped in this field seems to open the doors to a good career in the future. There are many different job areas where cybersecurity experts can work. The importance of cybersecurity is increasing today as companies and institutions digitize all their transactions. Government security units, banks and financial institutions, public institutions, large companies with databases and network providers have to employ these experts.

13. Prompt engineering

Prompt engineering is at the top of the list of professions with a bright future. This profession is the discipline of giving instructions to the model to obtain the desired output from arti-

126

ficial intelligence tools. In our language, concepts such as input or prompt engineering can be used for this profession. Engineers who enter prompts in a way that artificial intelligence understands aim to provide resources on certain issues. In addition to the prompt process, a good prompt engineer must have knowledge of the capabilities and limitations of artificial intelligence tools. At the same time, testing and evaluating model outputs and working to improve model performance are among the responsibilities of engineers. We can see that prompt engineers will work in collaboration with data scientists and product managers in the near future. Engineers must have skills in areas such as machine learning, natural language processing and programming.

14. Digital content creation

Digital content producers, among the newest professions, are people who produce content in a virtual environment. Although they have similarities with influencers, they also have differences. Content producers produce blogs; social media posts or website content in the field they are interested in. We can say that the most used platforms in this field are YouTube, Instagram, Twitter and TikTok. People address their followers on the internet with content such as text, video and audio. Digital content producers prepare attractive content suitable for their target audience. People who enrich the content with photos, videos and various visuals then publish their content on various platforms. It is important for those who want to build a career in this business line to follow digital media trends and innovations closely. In addition, it is important to regularly follow the performance of the content and update their content strategies. We can say that digital content producers can work as freelancers or find jobs under the roof of institutions such as agencies and brands. To be a successful content producer, it is necessary to follow social media trends and have strong communication skills.

15. Cloud security expertise

In recent years, cloud security expertise has become prominent and stands out among the most popular professions. People who deal with companies' data and systems in the cloud environment protect them against cyber attacks. For this, experts need to regularly analyze the company's cloud infrastructure and systems. In this way, experts can take precautions against cyber attacks. People who create security policies and procedures that are appropriate for the needs aim to ensure the security of cloud assets. Experts ensure the security of the system by building methods such as firewall, encryption, and authentication in the system. In addition, it is essential for those working in the field of cloud security to be familiar with the types of cyber attacks and security incidents and to produce solutions in possible attacks. We can state that it is important for people who are responsible for raising employees' awareness of cyber risks to follow new threats in terms of security. Those who want to make a career in this field need to have information about popular cloud computing platforms and to know security tools. In addition, it is important for experts to have the ability to analyze and produce solutions.

16. AI ethics expertise

The development of artificial intelligence has paved the way for the emergence of new branches of work in this field. One of these is the artificial intelligence ethicist. With the use of artificial intelligence, legal issues such as the protection of personal data have become especially important. In addition, it is obvious that artificial intelligence needs to be developed ethically in the future on issues such as transparency, openness, cyber attacks, and data privacy. Artificial intelligence ethicists, who will mediate the correct use of artificial intelligence technology, examine ethical and moral problems and produce solutions. The main duties of experts include examining and analyzing the ethical consequences of artificial intelligence tools. Individuals who evaluate the benefits and risks of artificial intelligence applications check how the tools work according to equality, privacy, and human rights. As a result of ethical analyses, experts establish basic principles for the responsible development of tools. It is expected that these principles will mediate the development of new policies by governments and developers. People who want to work in the field of artificial intelligence ethics expertise; It is necessary to be equipped in the areas of analytical thinking, problem solving, communication, artificial intelligence tools and the concept of ethics.

17. Internet of things (IoT) engineering

Among the new professions, IoT engineering deals with tasks such as connecting objects to the internet and collecting data. Engineers who work with various software, sensors and data transfer devices focus on developing IoT devices. People who build networks between the device and the internet ensure that the system operates securely. In addition, engineers examine the data obtained from the devices and use them to optimize business processes. People who constantly test the devices aim to minimize errors. In addition, we can say that engineers produce security solutions to protect devices against cyber attacks. In order to become an IoT engineer, it is essential to improve yourself in the field of hardware and software and to be proficient in network engineering. In addition, it is important for people to have strong problem-solving and data analysis skills.

18. Smart home design

Another job on the list of future career suggestions is smart home design. This profession, which emerged with the development of modern home systems, has become increasingly popular with the increase in smart homes. The main duties of smart home designers include finding solutions that will increase home comfort. Designers who analyze the demands and needs of customers help design smart home systems. In design; issues such as lighting, cooling, heating, security and entertainment are at the forefront. People who select the necessary equipment in line with customer requests, perform system installation and control. Designers who program the systems correctly also provide information about the system to the household. Those who want to make a career in this field must be an engineering or information technology graduate, familiar with smart home systems and have advanced analysis skills.

19. AI-supported urban planning

With the rise of artificial intelligence, AI-supported urban planning is among the professions that will become popular in the future. Planners who include AI tools in the urban planning process ensure that cities become more livable. Experts analyze the current status of the city by analyzing data from AI tools and sensors. This information paves the way for improvements in areas such as transportation, infrastructure development, buildings, land use, and emergency planning. Experts are expected to work in the future to produce solutions to issues such as city traffic flow and the development of public transportation systems with AI tools. In addition, constructing green buildings and using energy-efficient systems for a more efficient city are also among the duties of experts. People who will work in this field must be equipped in subjects such as artificial intelligence, engineering, and data science.

UNDISCOVERED NEW PROFESSIONS

Undiscovered professions represent new areas of work that will emerge due to future social, technological and environmental changes. These professions will take shape to meet needs that do not exist now, to use advances in technology and science, and to keep up with the changing dynamics of the world. Here are some professions that may emerge in the future and have not yet been discovered:

1.Human-computer interaction specialist

Explanation: As AI and human interaction become increasingly complex, experts will be needed to enable humans to communicate with computers in more efficient, intuitive, and natural ways. These experts can develop technologies that blur the boundaries between humans and AI.

Potential: Optimizing people's interaction with digital environments and increasing efficiency in all areas of daily life.

Genetic data consultant

Explanation: With the advancement of genetic engineering and biotechnology, a world where individuals' genetic data will provide more information about the factors that affect their health will be possible. Genetic data consultants will be able to offer health, lifestyle and nutrition advice based on the genetic makeup of individuals.

Potential: Personalized approaches to personal health management, disease prevention and treatment.

2. Mars settlement planner

Explanation: Space living will be further explored as a future possibility. As humans begin to establish colonies on Mars and other planets, settlement planners will design these new habitats.

Potential: To create sustainable living conditions in space and make Mars a place where humans can live.

Digital identity manager

Explanation: Digital identities are gaining importance as a new concept that helps individuals manage their online presence. Digital identity managers will be tasked with protecting personal data, ensuring digital security, and verifying individuals' digital identities.

Potential: Protecting against identity theft on the Internet, regulating and securing digital assets.

3. Human-machine compatibility expert

Explanation: In environments where people work together with robots and artificial intelligence, these experts will develop ergonomic and efficient solutions so that people can work in harmony with machines.

Potential: Ensuring that employees work together with robots, artificial intelligence and automation systems.

4. Water resources reuse expertise

Explanation: Global water scarcity and climate change will make efficient use and recycling of water essential. Water resources reuse experts will work on methods to clean wastewater and make it reusable and to use water efficiently.

Potential: Finding solutions to the global water crisis and ensuring sustainable water use.

5. Digital sociologist

Explanation: Sociologists who will examine human behavior in the digital world will investigate the effects of the internet, social media platforms and digital interactions on society.

Potential: In-depth analysis on the effects of social media, the structure of digital society and virtual relationships between people.

6. Climate expertise

Explanation: Climate change threatens ecosystems and life on Earth. Climate experts will develop innovative solutions to reverse or slow down climate change.

Potential: Managing weather conditions, restoring ecosystems, creating sustainable energy solutions.

7. Inner health coaching

Explanation: A coaching role to improve the mental, emotional and physical health of individuals. Inner health coaches will guide people on stress management, emotional balance and improving their overall quality of life.

Potential: Working in depth on personal development, mental health and overall well-being.

8. High-tech agricultural expertise

Explanation: With the combination of technology and agricultural methods; agricultural experts will now use biotechnology, drone technology and artificial intelligence to perform efficient and sustainable agriculture.

Potential: Developing innovative solutions that will revolutionize agriculture and increase efficiency in food production.

9. Mental health technology development

Explanation: Experts in this field can be defined as those who will develop digital solutions for psychological health issues. Mental health applications, artificial intelligence-supported therapists, virtual therapy platforms can be examples of this field.

Potential: Solving mental health issues with technology.

10. Fashion technology designer

Explanation: The fashion industry will evolve with wearable technologies, smart fabrics and eco-friendly production techniques. Fashion tech designers will ensure that high-tech clothing is functional and sustainable.

Potential: Combining fashion and technology to make clothes not only aesthetic but also practical and functional. The future workforce will be shaped by developing technologies, social changes and global issues. These professions are just a few of the areas that may emerge to provide solutions to social, environmental and economic challenges. The emergence of new professions will be shaped by opportunities and needs that have never been considered before.

THE MOST POPULAR HIGH EARNING PROFESSIONS

Today, the highest-paying professions in Turkey and the world vary according to developing technology, economy and the needs of the sectors. Here are some of the currently popular and high-paying professions:

Technology and software

- Artificial intelligence (AI) engineering
- Data science and analytics expertise
- Software engineering
- Cybersecurity expertise
- Blockchain development

Health and biotechnology

- Surgery (especially brain and heart surgery)
- Dentistry
- Aesthetic and plastic surgery
- Biotechnology expertise
- Psychiatry and psychology

Finance and investment

- Investment consultancy
- Portfolio management

- Risk analyst
- Cryptocurrency expertise
- Actuarial

Engineering and industry

- Artificial intelligence and robotics engineering
- Petroleum and natural gas engineering
- Electrical and electronics engineering
- Renewable energy expertise

Creative and media industry

- Digital content creation (YouTuber, TikToker)
- Film and series production
- Game design and development
- Graphic design and animation

Other popular professions

- E-commerce entrepreneurship
- International trade expertise
- Law (international and technology law)
- Logistics and supply chain management
- Management consultancy

Professionals who adapt to technology, are open to continuous learning and work for global markets generally earn higher incomes. In Turkey, software and digital services, the health sector and entrepreneurship are particularly noteworthy.

PROFESSIONS THAT WILL NEVER DISAPPEAR

Some professions will always continue to exist both in Turkey and in the world, despite technological developments and changing economic conditions. Because these professions are human-centered and it is difficult for them to be completely replaced by machines. Here are some of the professions that will never disappear:

Health sector

- Medicine (especially surgery and emergency medicine)
- Nursing and patient care
- Psychology and psychiatry
- Physiotherapy
- Pharmacy

Education sector

- Teaching (especially special talent and guidance classes)
- Educational consultancy
- Child development expertise

Food and agriculture sector

- Farming and agriculture expertise
- Food engineering

- Chef and cooking

Technology and engineering

- Software development (including artificial intelligence and robotics)

- Cybersecurity expertise

- Electrical and energy engineering

- Data analytics

Service and maintenance sector

- Security guards

- Cleaning and maintenance services

- Cargo and logistics workers

Art and creative works

- Artists and designers

- Film and TV series producers

- Interior designers

Law and finance

- Law (especially technology and international law)

- Financial consultancy

- Tax consultancy

Construction and infrastructure

- Civil engineering

- Architects

- Plumbers and technical service specialists

People-oriented professions

- Social workers

- Elderly care specialists

- Event and event managers

These professions, which require human relations, creative thinking and decision making, will always continue to exist despite technology.

In the future, promising professions are shaped by technological developments, social needs and new trends. Here are some of the prominent professions:

1. Technology and software

- Artificial intelligence engineering
- Data science and data analytics
- Cybersecurity expertise
- Blockchain expertise
- Developer (Web, Mobile, AR/VR)

2. Health and biotechnology

- Biomedical engineering
- Genetic engineering
- Robotic surgery expertise
- Digital health consultancy
- Elderly care and geriatrics expertise

3. Energy and environment

- Renewable energy engineering
- Energy efficiency consultancy
- Environmental engineering and sustainability expertise

4. Education and personal development

- E-learning experts
- Personal development and life coaching
- Educational technology developers

5. Design and creative sectors

• UX/UI designers

• Digital content creators

• Game developers

• 3D modeling and animation experts

6. Finance and economy

• Cryptocurrency and digital asset investment advisors

• Financial technology (fintech) experts

• Sustainable finance advisors

7. Logistics and supply chain

• Logistics technology experts

• E-commerce and supply chain managers

8. Social sciences and psychology

• Behavioral science experts

• Digital psychologists

• Media and communication consultants

Gaining competence in these areas and learning continuously will be important to survive in the changing world conditions. Which field interests you?

PROFESSIONS THAT WILL CHANGE

Yes, bakers, pastry chefs, chefs, restaurants and cafes will continue to exist in the future, but these professions and sectors will undergo major changes. Factors such as developing technology, changing consumer habits, sustainability and health trends will shape these sectors. Here are some predictions about how these sectors will evolve in the future:

1. Bakers and pastry chefs

• Technological advances: Increased automation and robotic technology in industrial bakery can speed up production processes. However, as long as traditional handmade products and organic ingredients are in demand, small bakers and boutique pastry shops can still survive.

• Consumer preferences: As interest in healthy, gluten-free, vegan or organic products increases, bakers and pastry chefs will continue to offer innovative products based on these demands. Additionally, desserts and breads customized to individual dietary needs may be popular.

2. Dessert shops

• Digitalization: Dessert shops will be able to reach wider audiences thanks to social media platforms and e-commerce. Online ordering and delivery services will create an important market for dessert shops.

• Innovative flavors: Different and creative flavors, fusion cuisine and the combination of international flavors will attract attention among young consumers. Desserts can continue to be prepared with healthier and more natural ingredients.

3. Chefs

• Compatibility with new culinary trends: Chefs will not only prepare traditional dishes, but also menus that emphasize healthy, sustainable and local products. In addition, meat alternatives, vegan dishes and dishes made with ingredients that have a lower carbon footprint may be popular.

• Training and expertise: In the future, chefs may become more specialized and offer personalized meals. For example, chefs with nutritional expertise may prepare health-focused dishes.

4. Restaurants and cafes

• Technological transformation: Restaurants and cafes will be equipped with systems that digitize order taking and will use artificial intelligence and robotic systems to enhance the customer experience. For example, robot waiters, automated kitchen equipment or online table reservation systems may become widespread.

• Social areas: Cafes and restaurants may cease to be places where only people eat and become social meeting areas, work areas or venues for cultural events. People can share on social media, hold online meetings or creative activities while eating.

• Health and sustainability: Restaurants and cafes will offer healthier menus to adapt to health trends. Organic, natural, low-calorie and locally produced ingredients will play an important role in line with customer demands. In addition, sustainable packaging and waste management will also be critical issues in the future.

5. Food and beverage export

• Globalizing markets: Thanks to online platforms and globalizing markets, restaurants and dessert shops can find opportunities to open up to international markets. This allows food cultures to be shared more globally.

• Artificial and alternative foods: The use of alternative resources in food production, especially innovative products such as artificial meats, plant-based proteins and milk alternatives, can enter the menus of restaurants and cafes.

6. Fast food and delivery services

• Food delivery technologies: In the future, food delivery services will become faster and more efficient. Drone and robot technologies can help deliver food to customers faster.

• Virtual restaurants: Instead of traditional venues, "art

restaurants" that only take online orders may become widespread. Such restaurants will not have a physical location, but will take orders by offering a menu through online platforms.

Bakers, pastry chefs, chefs, restaurants and cafes will evolve in line with technological developments in the future. While automation and digitalization increase efficiency, people will still value traditional flavors and handmade products. Themes such as personalized meals, healthy eating options and sustainability will play an important role in the future of these sectors. It will also be inevitable that businesses in these sectors will develop more creative and integrated solutions to improve the customer experience.

New professions and crafts

In addition to the professions that will disappear in the future, some new professions and crafts will emerge with technological developments and social changes.

1. AI and robotics experts

• People specializing in artificial intelligence, machine learning and robotics technologies will be in great demand in the future.

2. Virtual reality (VR) and augmented reality (AR) designers

• The number of experts working in the fields of virtual reality and augmented reality will increase, especially in the gaming, education and entertainment sectors.

3. Data Scientists and Data Analysts

• Data analysis and management will gain great importance in the digital world. The demand for professionals specialized in data scientists, algorithms and artificial intelligence systems will increase.

4. Sustainable energy experts

• Renewable energy sources and environmental sustainability will become a major sector in the future. Specialized crafts and professions will emerge in this field.

5. Content creation and digital marketing experts

• Professions such as digital media, influencers and content creation will become even more popular in the future with the power of social media.

6. Biotechnology and genetic engineers

• Genetic engineering, biotechnology and personal health data analysis will be important professions in the future.

With the rapid development of technology and social changes, some traditional professions and crafts will disappear or decrease significantly. However, new professions and crafts will emerge in areas such as digitalization, sustainability, biotechnology and artificial intelligence. In the future, adaptability and continuous learning will be the keys to success.

Most popular professions

Being an artist and being an athlete are both very popular and prestigious professions, depending on certain conditions and talents. However, success and popularity in these professions mostly depend on factors such as individual talent, marketing strategies, timing and luck. Although artists and athletes appeal to a wide audience, both professions are fields where there is intense competition and financial difficulties can be experienced.

The most profitable crafts

• **Watchmaking:** Craftsmen who provide services to luxury watch brands in particular earn high salaries and incomes.

• **Yemeni and leather processing:** Craftsmen who produce high-quality handmade leather shoes and bags can earn high incomes in luxury markets.

• **Coppersmithing:** Craftsmen who work with copper, especially in the antique and luxury kitchenware sectors, can earn high incomes.

• **Traditional carpet weaving:** Hand-woven carpets gain high prices, especially when sold to private collectors and luxury interiors.

• **Mother-of-pearl craftsmanship:** Especially traditional handcrafted furniture ornaments and luxury giftware manufacturers are profitable.

• **Fruit and vegetable carving:** Processing natural products, especially handcrafted, can provide high incomes in the gift sector.

Today, high-paying professions generally consist of jobs that require expertise, skill and experience. Many years of education and experience are essential to be successful in these professions. Crafts can also generate high incomes in some niche areas, but this often depends on the quality, market, and value of the craftsmanship.

Less known traditional professions in Turkey and the World

Little-known or forgotten handicrafts and traditional professions, both in the world and in Turkey, are an important part of cultural heritage. Here are some notable examples:

Lesser known handicrafts and traditional professions in Turkey

Anatolia, with its rich cultural heritage, has hosted many traditional arts. However, due to modernization and technological developments, some handicrafts and crafts have been forgotten. Here are the arts that are beginning to disappear or are in danger of disappearing in Anatolia:

1. Wire-breaking (Bartın): It is an elegant decorative art made by processing thin metal wires onto silk or fabric. It has a deep-rooted history especially in the Bartın region.

2. Mother-of-pearl inlay (Gaziantep, Kahramanmaraş): It is the art of creating decorative patterns by processing mother-of-pearl pieces onto wooden surfaces. Examples can be seen especially in Topkapı Palace. The number of masters has decreased over time.

3. Felting (Afyon, Konya, Manisa): Felting is a traditional craft made by turning natural wool into felt and turning it into clothing, accessories or decorative products, and by wetting and beating the wool into felt. In the past, it was important for shepherd's clothes and blankets.

4. Yemeni making (Gaziantep, Kahramanmaraş): Yemenis, which are handmade leather shoes, are produced by a small number of masters today.

5. Pottery (Avanos, Kütahya, Manisa): Pottery, which is made in many parts of Anatolia, especially in Avanos, has lost value with the increase in factory-made ceramics.

6. Needle lace (Bursa, Izmir, Denizli): This lace art, which requires craftsmanship and in which very fine and elegant lace motifs are made with needles, carries traditional Anatolian motifs.

7. Wooden spoon making (Konya, Beyşehir): Handmade wooden spoons have lost their importance with the spread of plastic and metal kitchenware.

8. Savat work (Van): It is an ornamental art made with black alloys on silver. It is specific to the Van region and is continued by a small number of masters.

9. Tile making (Kütahya, İznik): The art of patterning on handmade ceramics is traditional, especially in Kütahya and İznik. Made tiles are in danger of being forgotten due to modern ceramic production.

10. Basket making (Black Sea, Aegean): Handmade baskets woven from natural plant fibers have become rare.

11. Kilim weaving (Şanlıurfa, Isparta, Uşak): Hand-woven kilims have lost their popularity in the face of factory-made carpets.

12. Wood carving art (Safranbolu, Trabzon): It is one of the traditional Turkish handicrafts used in furniture and decorative items. Although its embroidery is used especially in mosque decorations and furniture, the number of masters is rapidly decreasing.

13. Mohair weaving (Ankara): Mohair fabric obtained from the wool of the Ankara goat was used for luxury textile products in the past, but is rarely produced today.

14. Kutnu weaving (Gaziantep): Traditional fabric production, woven from a mixture of silk and cotton threads, is carried on by very few masters today.

15. Cane making (Devrek, Zonguldak): Canes made by hand are famous for their decorations and motifs. However, the number of masters has decreased.

16. Ebruzenlik (marbling art): This art, which is done by transferring the patterns created on the surface of water to paper, requires mastery.

17. Katı' art: This traditional art, known as paper carving art, is done with great patience and precision.

18. Stone doll making (sille dolls): The making of stone dolls specific to the Sille village of Konya is traditional.

These arts reflect the cultural memory of Anatolia and are in need of protection.

Lesser known crafts and professions in the world

1. Kintsugi (Japan): The art of repairing broken ceramics with gold or silver powder.

2. Shibori (Japan): The technique of knotting fabrics with special methods and patterning them with natural dyes.

3. Pietra Dura (Italy): The art of mosaic decoration made with semi-precious stones.

4. Scrimshaw (North America): The art of carving on whalebone or teeth.

5. Tjanting Batik (Indonesia): The art of creating patterns on fabric with beeswax and coloring them with natural dyes.

6. Tanka (Tibet): The art of painting on fabric made for religious purposes and including fine embroidery.

7. Zardozi (India): The art of embroidery made with gold and silver threads.

8. Filigree (Portugal): Intricate jewelry designs made with thin gold or silver wires.

9. Matryoshka doll making (Russia): The painting and design of handmade wooden dolls that fit together.

10. Mola Technique (Panama): The Kuna Indian art of embroidering brightly colored layered fabric.

Some of these arts are in danger of extinction.

NEVER ENDING CRAFTS

Traditional and handicraft professions maintain their existence in many parts of the world because they are the carriers of cultural heritage. Despite developing technology, these professions do not lose their importance in line with both artistic and practical needs. Here are some of the handicrafts, traditional and deep-rooted professions that will never end in Türkiye and the world:

Traditional professions and handicrafts that will never end in Turkey

Handicrafts

• **Tile making:** Traditional handcrafted tile production in Kütahya and İznik

• **Marbling art:** The art of transferring patterns onto paper by drawing on water

• **Calligraphy: Traditional Ottoman calligraphy**

• **Coppersmithing:** Widespread in cities such as Gaziantep and Kahramanmaraş

• **Wood carving:** Furniture and ornamental items production in regions such as Trabzon and Kastamonu

• **Felting:** Handicrafts in cities such as Denizli and Afyonkarahisar

• **Weaving and carpet making:** Famous carpets in Hereke, Isparta, Uşak and Manisa

• **Filigree (silversmithing):** Traditional silver art in Mardin and Midyat

Traditional professions

• **Saddlery:** Production of saddles and saddles for cargo vehicles.

• **Clogs:** Production of traditional wooden slippers.

• **Quilting:** Production of cotton and wool quilts by hand.

• **Saddlery:** Production of leather belts, bags and riding gear.

• **Knifemaking:** Production of knives famous in Bursa and Sivas.

Traditional professions and handicrafts that will never end in the world

Handicrafts

• Murano glasswork (Italy)

• Origami and calligraphy (Japan)

• Kilim weaving (Iran, Morocco)

• Wood carving (Nordic Countries)

• Ceramic embroidery (Spain)

• Indian textile painting (Bandhani, Batik)

• South American silversmithing (Peru, Argentina)

Traditional professions

• Shoemaking (Italy)

• Watchmaking (Switzerland)

• Winemaking (France)

• Shipbuilding (Norway)

• Artisanal cheesemaking (Netherlands)

Common characteristics of never ending professions

- It is the carrier of cultural heritage.

- It requires manual labor and mastery.

- It is focused on personal design and art.

- It cannot be replaced one-to-one with technology.

- They maintain their value thanks to the increasing global demand for handmade products.

These professions will continue to maintain their importance both in Turkey and in the world, as they are supported by tourism and the art economy. Traditional handicrafts and deep-rooted professions in Turkey are important parts of the rich cultural heritage. Despite technological developments, these branches of art and crafts continue to exist thanks to both tourism and cultural value. Here are some of the traditional handicrafts and crafts that will never end in Turkey:

1. Weaving and textile arts

- **Carpet and kilim weaving:** Famous in Isparta, Uşak, Hereke, Sivas and Manisa regions.

- **Felting:** Handmade felts are produced in cities such as Denizli and Afyonkarahisar.

- **Yazma (head scarf) and printing:** Traditional in Tokat and Kastamonu.

2. Woodworking

- **Wood carving:** It is widespread in the Trabzon, Kastamonu and Safranbolu regions in furniture and ornamental items making.

- **Clogs:** Traditional Ottoman bath slippers making (Bursa and Hatay).

3. Ceramics and tile making

- **Iznik tile work:** Traditional ceramic and tile art made in Iznik and Kutahya.

- **Earthenware work:** Famous in Nevsehir and Avanos.

4. Metalworking

• **Coppersmithing:** Copper pots, trays and ornaments are produced in Gaziantep, Kahramanmaraş and Diyarbakır.

• **Filigree:** Fine silversmithing in Mardin and Midyat.

• **Tinsmithing:** Cleaning and polishing of copper kitchenware (Sivas and Tokat).

5. Glass arts

• Glass blowing and bead making: Nazarköy village of Izmir is known for making evil eye beads.

• Glass carving: Traditional glass processing arts.

6. Leather and saddlery work

• **Saddlery:** Leather bag, belt and saddle making (Bursa and Afyonkarahisar).

• **Shoemaking:** Hand-made shoe making (Gaziantep).

7. Jewelry and stone work

• **Oltu stone workmanship:** Prayer beads and jewelry made with oltu stone in Erzurum.

• **Emerald and ornamental stone workmanship:** Widespread in Cappadocia and Mardin.

8. Paper and writing arts

• **Marbling art:** Traditional Turkish art of creating surface patterns.

• **Calligraphy art:** An important art of writing that has been carried from the Ottoman Empire to the present day.

9. Food and gastronomy crafts

• **Turkish delight and candy making:** Famous in Afyonkarahisar and Istanbul.

• **Pastirma and sausage making:** Traditional in Kayseri.

10. Other traditional arts and crafts

• **Quilting:** Making hand-embroidered quilts (Bursa and Denizli).

• **Saddle making:** Making saddles for pack animals (Balikesir and Tokat).

• **Pottery:** Traditional in Nevşehir (Avanos) and Kütahya.

WHAT SHOULD UNEMPLOYED PEOPLE DO?

In the future, rapid developments in technology will lead people to new life and work models as automation and artificial intelligence transform or completely eliminate many occupational groups. Individuals who become unemployed will have to develop different strategies to adapt to this transformation. Here are the scenarios that may occur during this process:

- **New professions and sectors**

Technology will create entirely new professions while destroying some jobs.

- **Artificial intelligence and robotics expertise:** Fields such as robotics technicians, data analysts, and artificial intelligence training will become more popular.

- **Digital creators:** Digital professions such as virtual reality designers, content producers, and augmented reality experts will become widespread.

- **Sustainability sector:** Renewable energy, environmental engineering, and circular economy expertise will be in high demand.

- **Continuous training and skills transformation (reskilling)**

Continuing education programs and skill development processes will become important for unemployed individuals.

- **Online education platforms:** People will be able to gain professional skills from platforms such as Coursera and Udemy.

- **State-supported trainings:** To make the workforce tech-savvy, governments and companies can offer training programs.

- **Creative economy and entrepreneurship**

With the tools provided by technology, individuals can start their own ventures or develop creative business models.

- **Freelance economy:** Digital platforms for freelancers (such as Upwork, Fiverr) will grow.

- **Small startups:** The number of individuals offering products/services by establishing their own e-commerce platforms will increase.

- **Universal basic income - UBI**

- **State-supported income:** With the efficiency gains provided by robots and automation, governments can offer universal income to individuals to meet their basic needs.

- **Trials:** UBI has been trialled in Finland, Canada and some US states; the system may become widespread in the future.

Social service and community participation

Unemployed individuals can be directed to social service projects to contribute to society.

- **Voluntary programs:** Active roles can be undertaken in social solidarity projects.

- **Community education and mentoring:** Experienced individuals can guide young people.

Personal development and hobby economy

- **Hobby income:** Craft production, art, music and other hobbies can become sources of income.

- **Health and lifestyle areas:** Professions such as yoga teaching and health coaching may increase.

Transformation of social security and welfare systems

• **New social policies:** Unemployment benefit systems can be transformed and different economic solutions can be offered to individuals to sustain their lives.

• **Conclusion**

As technology transforms the world of work, individuals will need to be flexible, learn new skills, and develop creative solutions to adapt. Societies and governments will also need to develop policies appropriate for this transformation and support individuals who are unemployed.

FLEXIBLE WORKING HOURS, PEACEFUL ENVIRONMENT

Professions and crafts that provide a happy and peaceful environment while working are generally found in jobs where human interaction is positive, stress levels are low, and personal satisfaction and creativity are at the forefront. In addition, flexible working hours, creative freedom, and meaningful work that increase the employee's quality of life and job satisfaction also fall into this category. Here are such professions and crafts:

Therapists and psychologists

Why is it peaceful and satisfying? Helping people and improving their quality of life is deeply meaningful to many therapists and psychologists. This profession is a line of work that provides the satisfaction of helping others.

Work environment: Therapy rooms are usually calm and reassuring. They may work in their own offices or in hospitals or private clinics.

Artists (painting, sculture, photography)

Why is it peaceful and satisfying? The creative process is both a source of therapy and peace for many artists. Creating art provides emotional expression and personal satisfaction.

Working environment: Artists usually work in their own studios, in a quiet and comfortable environment. Personal space and freedom are very important while working.

Writers and editors

• **Why is it peaceful and satisfying?** Writing and editing are professions that provide freedom to express one's thoughts and creativity. Working on your own projects or directing the projects of others can be satisfying.

• **Work environment:** Most writers and editors prefer to work in a comfortable environment, choosing their own workspace. Spending time in a quiet and creative environment can be peaceful.

Educators (teachers, trainers, mentors)

• **Why is it peaceful and satisfying?** Teaching offers the opportunity to inspire and help people develop while imparting knowledge. Students' success and development are a great source of motivation for educators.

• **Work environment:** Although classroom environments can be intense at times, teachers generally derive great satisfaction from the relationships they establish with students and their success. The work environment is shaped by the teacher's style, but establishing positive relationships with people makes this profession satisfying.

Gardeners and landscape designers

• **Why is it peaceful and satisfying?** Being in touch with nature and working with plants is a stress-relieving and peaceful activity. Gardening and landscape design require creative thinking and a connection with nature.

• **Work environment:** Usually working in an open-air environment, surrounded by nature. The work area is surrounded by the calming effects of nature.

Yoga and meditation instructors

• **Why is it peaceful and fulfilling?** Yoga and meditation instructors help people reduce stress and improve their mental and physical health. They work in an environment where they find inner peace.

- **Work environment:** Yoga studios and meditation spaces are often created to be calm and peaceful. Instructors guide people in a calm atmosphere

Pet sitters and vetenerians

- **Why is it peaceful and fulfilling?** Working with animals can be a fulfilling and peaceful profession for many people. Helping animals heal or care for them requires love and compassion.

- **Work environment:** Time spent with animals has a direct positive impact. Veterinarians and caregivers may work in clinics or in open areas.

Cooks and pastry chefs

- **Why is it peaceful and satisfying?** Cooking is an art form for many people, and serving delicious food to others can be fulfilling. This profession involves being creative and producing something.

- **Work environment:** Chefs can often work in a relaxed environment in the kitchen. Many chefs create a creative and peaceful atmosphere in their kitchens.

Interior designers

- **Why is it peaceful and satisfying?** Interior designers create aesthetic and functional spaces by using their creativity while beautifying people's living spaces. Meeting the needs of their clients is one of the jobs that give them satisfaction.

- **Work environment:** Designers can work in offices where they will carry out their projects or directly in their clients' homes. The work environment is usually calm and orderly.

Handicrafts and craftsmen (especially jewelry and fashion design)

- **Why is it peaceful and satisfying?** Handicrafts are a line of work that combines creativity and elegance. Works such as handmade jewelry design, leather work, and furniture making are works that have aesthetic value and provide personal satisfaction.

- **Work environment:** Craftsmen usually carry out their creative processes in their own workshops in a calm atmosphere.

Music and performance artists

• **Why is it peaceful and satisfying?** Making music is a powerful way to express and share emotion. Artists can experience a deep sense of fulfillment when they express their inner world on stage.

- **Work environment:** Concerts or performances bring artists together with people in a lively environment. The stage environment can be a source of happiness and peace for many artists.

Professions that work in a happy and peaceful environment while working are generally professions that aim to do meaningful work for people, offer creative freedom, and provide personal satisfaction. Those working in these professions find emotional satisfaction while also making positive contributions to the people around them. Whether it is being in touch with nature or guiding others, a peaceful work environment usually has a positive psychological effect.

PROFESSIONS THAT CAN BE DONE WITHOUT

A DIPLOMA

There are many jobs that can be done without studying, that is, without a university degree. These are usually jobs that focus on skills, experience, entrepreneurship and talent. Here are some examples:

Art and handicrafts

- Hairdressing and barbering
- Tattoo artistry
- Jewelry design
- Tailoring
- Furniture and woodwork
- Bakery
- Cooking

Business and entrepreneurship

- E-commerce sales
- Marketing
- Store or cafe management
- Brokerage and real estate consultancy

Technique and craft

- Electrical and plumbing
- Welding
- Automotive repair and tire repair
- Painting and decoration

Service sector

- Cargo and courier services
- Driving (taxi, shuttle, truck)
- Cleaning services
- Security guard

Digital and creative jobs

- Graphic design (can be learned through courses)
- Social media management
- Photography and video production
- YouTube content creation

Agriculture and animal husbandry

- Farming
- Beekeeping
- Animal Husbandry

Other professions

- Organization and event management
- Market and fair stand management
- Restaurant or buffet management

Instead of a diploma, skills, experience and customer satisfaction are important factors for success in these professions. You can also increase your competencies with various courses and certificate programs.

The most interesting professions

The most interesting professions, crafts and jobs in the world are often those that require extraordinary skills, are unique or unusual, and are sometimes in intriguing areas. Sometimes these professions have a historical past, while other times they have emerged from the creative or innovative fields of modern times. Here are some interesting professions and crafts from around the world:

1. Storm chaser: Storm chasers are professionals who track and scientifically study hurricanes, tornadoes, and other extreme weather events. This profession is very interesting for both naturalists and adventure seekers. It is very exciting to encounter the power of nature while tracking storms and to collect scientific data about these events.

2. Star *chaser* or astronomical advisor: These people observe astronomical constellations and sky phenomena and guide people in this regard. They can also provide astrology-based consulting services. Being able to make both scientific observations and provide personal life guidance is a very different and creative field.

3. Professional *game tester*: Game testers play video games to find bugs and check if the games work properly. This can be a great career for those who love to be involved in the gaming world. Testing games and sometimes fixing bugs in old games can seem like a pretty fun job.

4. Sailboat inspector: They are experts in inspecting sailboats and yachts. This profession is carried out to evaluate the safety, functionality and design quality of boats. Being on the water and examining different boats provides a very interesting experience.

5. *Cryptozoologist*: It is the name given to scientists who investigate whether mythical creatures and animal species exist. This profession involves making discoveries to search for imaginary or mythical creatures. This profession arouses great curiosity in those who investigate many mysterious creatures and includes fantastic elements.

6. Ice cream sculptor: Ice cream sculpture artists. These people create original and innovative works of art for exhibitions and events by aesthetically shaping ice cream. Creating artistic creations with a delicate material like ice cream is a very challenging but fun task.

7. Wine taster (*sommelier*): Wine experts are professionals who determine the quality and variety of wines and which dishes they go best with. It requires deep knowledge of the world of wine and tasting expertise. This profession has also become a culture.

8. *Chocolate sculptor*: They are artists who create creative works with chocolate. They can create Christmas trees, statues or chocolate sculptures for special events. Since chocolate is a material that is easy to shape and melt, artists can overcome difficulties while achieving a sweet result.

9. *Recycled art creator*: They are people who create works of art from recycled materials. Sculptures, wall decorations and more can be made using waste such as plastic, metal, glass. In today's world where natural resources are limited, reusing waste is a very interesting and creative area from an artistic perspective.

10. *Treasure chest restorer*: They are experts in restoring antique chests and old boxes. These chests, which have both historical and aesthetic value, gain new value through the restoration process. Restoration of old items is a very interesting profession that combines history and art.

11. Frozen food specialist: hey are the people who produce and develop frozen food products. They specialize in the proper storage, quality control and development of frozen foods. Frozen food production is much more than an ordinary cooking profession that requires scientific and innovative techniques. Interesting professions and crafts that exist in the world and in Turkey are jobs that highlight the creative aspects of people, sometimes intertwined with nature and sometimes combining technology and art. While each requires unique skills, they can also be inspiring for those who follow their passions in different areas. Such professions offer a great opportunity for those looking for a different career path that is far from ordinary.

Difficult craftsmanship

Craftsmanship refers to specialization in jobs that require both technical knowledge and manual skills. Some craftsmanship professions require more experience, fine workmanship and attention than others. Here are some professions in the difficult craftsmanship category:

1. Filigree (silverwork): Filigree is a craft made by finely processing silver or gold wire to create patterns. It requires very fine workmanship and requires patience, precision and attention. It is quite difficult to work with detailed patterns and tiny pieces while doing filigree.

2. Glass blowing (glass artistry): Glass blowing is a craft made by shaping molten glass at high temperatures. Glass artists carry serious risks because they work at temperatures above 1000°C. This craft requires technical knowledge, experience and physical strength.

3. Classic wood carving: Wood carving requires working with especially fine details. Great skill is required to not damage the texture of the wood, to use the right tools and to create an aesthetic form. Hand carving is a difficult craft that requires patience and attention.

4. Leather Workmanship (Leather Master): Leather work involves shaping the leather, sewing and creating detailed patterns. Correct sewing and assembly techniques determine the quality of the work. Leather cutting, sewing and dyeing processes must be done carefully.

5. Miniature art: Miniature art involves making small-scale pictures and patterns. In this art form, very small details are processed with thin brushes. The artist goes through a process that requires observation skills and patience.

6. Tile and ceramic art: Tile and ceramic making involves a wide variety of techniques: pottery, ceramic firing, and pattern making. Craftsmen use their physical and creative skills to create a product that is both aesthetic and functional. In addition, processes such as firing and glazing ceramics require attention and care.

7. Fabric weaving and carpet weaving: Traditional hand-woven carpets and fabrics require great mastery. The orderly placement of threads and the correct weaving of patterns require great experience. In addition, carpet and fabric weaving is a time-consuming and patient process.

8. Jewelry and jewelry making: Jewelry making involves the shaping of jewelry made with gold, silver or precious stones. Jewelry making requires very fine workmanship and an understanding of aesthetics. In addition, the processing and assembly of precious stones are processes that require precision.

9. Steelworking and blacksmithing: Blacksmithing is the art of shaping and processing steel. The shaping of steel with heat requires working with heavy materials in particular. Craftsmen produce aesthetic and functional metals by working at high temperatures.

10. Instrument making: Making musical instruments is a job that requires both engineering and art. Traditional instruments in particular require a high level of skill to ensure that each part fits together correctly. Instruments such as guitars, pianos, and violins must be meticulously crafted for both sound and aesthetics.

11. Glass work (Stain glass or windows): Cutting, assembling and fixing colored glass is one of the basic processes of glassmaking. It requires accurate cutting of glass and patient assembly. It also requires great care in cutting and assembling the glass.

12. Game manufacturing (wooden toy making): Handmade toys are crafts where every detail is meticulously crafted.

Cutting, painting and assembling wooden toys correctly requires being a good craftsman. Safety and functionality should also be taken into consideration.

Conclusion

Difficult crafts often require a high level of attention to detail, fine workmanship, technical knowledge and experience. Patience and physical skills are often important. These crafts can take time to master, but producing high-quality products makes these crafts both creative and rewarding.

V

CITIES OF THE FUTURE

SMART CITIES

Smart cities are cities where digital technologies and data analytics are integrated to improve the infrastructure, services and quality of life of cities. These cities aim to create a more sustainable, efficient and livable environment by using data collection and analysis techniques. Smart cities aim to meet the needs of individuals and societies more effectively and are usually supported by technologies such as the internet of things (IoT), artificial intelligence, big data, sensors and communication networks.

Basic components of smart cities

- *Intelligent transportation systems*

Smart transportation uses digital technologies to optimize traffic flow within the city, prevent traffic accidents and make transportation infrastructure efficient.

- **Autonomous vehicles and shared transportation:** Public transport such as autonomous vehicles, taxis and buses can reduce traffic in cities. In addition, transportation becomes more flexible with shared car and bicycle systems.

- **Smart traffic lights:** Traffic lights use sensors and artificial intelligence to monitor traffic flow and adjust green light times to reduce traffic congestion during rush hour.

- **Instant traffic information and navigation:** In smart cities, traffic conditions can be monitored with instantly collected data and guidance can be provided to drivers. This helps drivers avoid traffic congestion and choose alternative routes.

- *Smart energy managament*

In smart cities, energy consumption is monitored and optimized through digital systems. This is critical for improving energy efficiency and reducing environmental impact.

- **Smart Grids:** Electricity networks monitor energy demands to distribute power more efficiently. They also optimize the flow of energy from renewable energy sources (solar, wind).

- **Energy storage systems:** Smart energy systems can store excess energy and use it during times of peak demand. This reduces energy waste and supports sustainable energy use.

- *Smart water and waste management*

Water and waste management are integrated with smart systems to ensure efficient use of resources and minimize environmental impacts.

- **Water consumption and distribution:** Sensors used in water networks monitor water losses and prevent unnecessary water use. In addition, water is distributed at the right time and amount.

- **Waste management:** Smart waste bins measure the filling level inside with sensors and only direct the waste collection vehicle when it is full. This saves both time and fuel.

- **Smart buildings and infrastructures**

Smart buildings use various technologies to increase energy efficiency and improve the quality of life.

- **Energy efficient buildings:** In smart buildings, heating, cooling, lighting and water usage systems are monitored and optimized with sensors. This saves energy and reduces the carbon footprint.

- **IoT connected structures:** IoT devices digitally monitor and manage every area of the building, automatically adjusting environmental conditions (temperature, humidity, light) according to the needs of the users.

- *Digital public services*

In smart cities, public services become accessible through digital platforms, making citizens' interactions with the government faster and more efficient.

- **E-government applications:** Citizens can apply for various public services, pay taxes and easily access municipal services over the internet.

- **Smart healthcare:** Smart health technologies provide remote monitoring and treatment services for patients in the city. In addition, more effective policies for public health can be created by analyzing health data.

Advantages of smart cities

- **Sustainability:** Smart cities minimize environmental impacts by using natural resources efficiently. Renewable energy systems, waste management and water resources management offer an environmentally friendly city life.

- **Efficiency and time saving:** Digital infrastructures reduce time loss by optimizing all processes in the city. Traffic congestion is reduced, public services are accelerated and resource use becomes more efficient.

- **Quality of life:** Smart cities are designed to improve the quality of life of their citizens. The quality of life in the city is increased with technologies that reduce traffic congestion, increase safety, minimize environmental impacts and facilitate healthcare services.

- **Economic opportunities:** Smart cities can become hubs of technology and innovation, paving the way for new business opportunities, entrepreneurship, and economic growth.

Challenges and ethical issues

- **Data security and privacy:** Sensors and data collection systems used in smart cities process large amounts of personal data. The security of this data and the privacy of citizens will be an important issue. Strong security measures are required against cyber attacks.

- **Digital access and inequality:** Not all citizens may have access to digital infrastructure. During the development of smart cities, it is important to ensure equal access so that the digital divide does not widen.

- **High costs:** Building and maintaining smart cities requires high costs, which can be a barrier, especially for developing countries.

- **Technological dependency:** In smart cities, all infrastructure and services are based on technologies. This can lead to system collapse and major disruptions in the event of a major failure or cyber attack.

Smart cities are shaping the cities of the future by integrating technologies into urban life. These cities are using digital technologies to increase efficiency, sustainability and quality of life. However, there are also significant challenges in establishing and maintaining such cities, such as security, equal access and ethical issues. In order for smart cities to be successful, planning should be done by paying attention to all these issues.

SHARED HOUSES, HOUSING

Housing, travel and lifestyle trends are changing rapidly. With technology, sustainability understanding and individualization demands, tiny houses, shared housing systems such as Airbnb, caravan life and different housing models will undergo major changes in the future.

Tiny house and small house trend

• **Minimalist life:** People are turning to smaller homes in search of less consumption and a simpler life.

• **Smart designs:** High-tech, energy-efficient, compact home designs will become popular.

• **Sustainable materials:** Tiny houses built with recyclable and environmentally friendly materials will become widespread.

• **Community spaces** Tiny house villages can thrive with common areas that encourage social interaction.

Artificial Airbnb and shared home models

• **More automation:** Automatic reservation and home management systems supported by artificial intelligence.

• **Decentralized platforms:** Blockchain-based platforms can make it possible to rent homes without any intermediaries.

• **Personalized experiments:** Automatic recommendations and personalized services will be provided based on the user's past preferences.

- **Virtual tour and AR technology:** Users will be able to tour homes in virtual reality before making a reservation.

- **Green certified homes:** Homes that use sustainable energy will be in higher demand.

Summer houses and holiday homes

- **Flexible ownership model:** With fractional ownership systems, more than one person will be able to use the same summer house.

- **Seasonal use:** Instead of summer houses, different locations will be preferred in different periods with flexible rental options.

- **Metaverse holiday homes:** Virtual summer houses can be purchased and rented on digital platforms.

Caravan and mobile living

- **Caravan technology:** High-tech caravans equipped with solar panels, smart energy management and IoT systems.

- **Mobile living communities:** Caravan parks can become social community centres.

- **Legal regulations:** Infrastructure and laws suitable for RV life may expand.

- **Environmentally friendly designs:** Electric or hydrogen fueled caravans that minimise their carbon footprint.

- **Subscription models:** Shared caravan rental platforms may become widespread.

General trends in future life patterns

- **Flexible housing use:** People will travel and work without being tied to a specific place (remote work + travel).

- **Virtual and physical integration:** Holiday homes or caravans will become integrated with virtual offices.

- **Community based living:** People will turn more towards shared living spaces for both economic and social reasons.

- **Green and sustainable housing:** Energy efficient, eco-friendly homes will become the norm.

In summary, people will adopt more mobile, flexible and sustainable lifestyles in the future. Tiny houses, caravan living and shared homes will be taken to new dimensions with the demands of technology, sustainability and individualization. Holiday and living experiences will no longer be just physical, but also virtual and community-oriented.

VI

FUTURE LIFESTYLE

HOUSE CLEANING AND DOMESTIC CARE

House cleaning

• Otonomous robots: Robot vacuum cleaners that are already in use will gain more advanced forms. These robots will not only sweep, but will also do tasks such as floor cleaning, dusting and window cleaning.

• Smart surfaces: Self-cleaning floors, walls and windows will become common. Special coatings will produce surfaces that do not retain dirt and stains.

• Air purifying systems: Air cleaning and disinfection systems that continuously optimize indoor air quality will become standard.

Cooking

• **Automatic food robots:** Robot chefs that automatically prepare and cook ingredients will become widespread. Users will be able to prepare meals by simply selecting a recipe and placing ingredients.

• **Smart ovens and pots:** Smart kitchen appliances will be used that automatically adjust the ideal cooking time and temperature of the food and warn before the food burns.

• **3D food printers:** 3D printers that print meals according to desired recipes will prepare meals suitable for personal diets.

Pet care

- **Smart feeders:** Devices that manage pets' feeding schedules and automatically meet their food and water needs will develop.

- **Health tracking systems:** Health problems can be diagnosed early thanks to devices that monitor pets' movements, sleep patterns and health conditions.

- **Automatic toilet systems:** Self-cleaning systems that meet pets' toilet needs will become standard.

Ironing and laundry

- **Wrinkle-free fabrics:** In the future, smart textile products that do not wrinkle and do not stain will become widespread, and the need for ironing will be largely eliminated.

- **Automatic folding robots:** Robots that fold and organize laundry after washing will become widespread.

- **Smart washing machines:** Machines that detect stains, select the appropriate washing program, and automatically perform processes such as drying and disinfecting will be used.

Washing up

- **Smart dishwasher robots:** Robotic systems that clean and place plates taken from the table will be developed.

- **Self-cleaning kitchenware:** In the future, pots, pans and plates will be self-cleaning.

Welcoming guests and home security

- **AI assistants:** Artificial intelligence-supported systems that recognize and greet guests at the door and send notifications to the host will become widespread

- **Smart home management systems:** Home lighting, temperature control and security systems will be fully automatic.

Laundry

- **Smart facric cleaning cabins:** Systems that clean and disinfect clothes without washing and with low energy consumption will be used.

- **Water saving systems:** Washing machines that recycle water and wash with minimal water will become standard.

In the future, housework will be carried out by automated systems that require minimal human labor. Homes will become high-tech living spaces designed to allow people to rest and spend more time on creative activities.

WILL WE HAVE A VIRTUAL FAMILY?

Yes, it is quite possible that the concept of a virtual or robotic family will emerge in the future. Technological developments, social changes and individual preferences can pave the way for such innovations. Here are the possible developments in this regard:

1. AI-supported "virtual family"

• Virtual families consisting of artificial intelligence members that will meet people's emotional needs may become popular.

• These family members can adapt to the person's life and take on roles such as talking, giving advice, and providing emotional support.

2. Robotic family members

• Physical robots at home can play child, spouse or parent-like roles.

• Especially for lonely individuals or the elderly, robotic assistants can provide both companionship and care services.

3. Digital holographic families

• Virtual family members can be created with hologram technology. People can interact with these holographic beings in a physical environment.

4. Metaverse families

• Environments can be developed in virtual worlds where users can establish family units with each other.

• In such families, people can take on the roles of "spouse," "child," or "friend" with their avatars.

5. Reduction of genetic and social pressures

• As the pressure to form a "real" family decreases in society, people may turn to virtual or robotic alternatives instead of traditional family structures.

6. Robotic children

• Robotic children, which have begun to emerge in Japan and some developed countries, may become an option for individuals who want to establish an emotional bond in the future.

• These robots will have the capacity to learn and will be able to adapt to their "parents" over time.

7. Ethical and social debates

• Ethical and moral debates may also arise against such family structures. The replacement of real human relationships with virtual ties may lead to criticism.

As a result, virtual and robotic families may emerge in the future as an alternative to the traditional family concept and meet the social needs of individuals in different ways.

HOW WILL FRIENDSHIP CHANGE?

In the future, the concepts of friendship and companionship may transform with the changes in technology and social dynamics. Here are the possible developments in this regard:

1. Digital and virtual friendships

• Friendships established through avatars in virtual worlds such as the Metaverse may become more common.

• People can develop meaningful friendships by meeting individuals from different cultures and geographies in virtual reality environments.

2. AI "friends"

• Artificial intelligence-supported virtual friends that can be adapted to personal needs can reduce people's feelings of loneliness.

• These artificial friends can learn users' interests and offer constantly evolving conversations and provide emotional support.

3. Friendship through common interests

• Advanced algorithms and artificial intelligence-supported platforms can enable stronger friendships to be established by matching people based on personality analysis and interests.

4. Physical and digital social communities

- Hybrid social events (both physical and digital) may become more common. For example, individuals viewing an art exhibition both on-site and virtually may interact.

5. Emotion detection technologies

- Thanks to wearable devices and biometric sensors, friends can more easily understand each other's moods, allowing for deeper and more empathetic relationships.

6. More inclusive and global relations

- As the importance of cultural, language and geographical boundaries decreases, people can communicate more easily with different societies.

7. Social ties independent of time and space

- Thanks to hologram technology, people will be able to chat as if they were in the same room, even if they are in different cities or countries.

8. Dating "subscription" platforms

- Perhaps in the future, people will join membership platforms for private social networking or dating experiences. These platforms can make it easier to build trustworthy, quality relationships.

9. Revaluation of human relations

- As the risk of becoming isolated through technology increases, face-to-face friendships and deep human connections may become a "luxury" and people may value these connections more.

These developments can both enrich the concepts of friendship and companionship and add new dimensions to the nature of human relations.

FAMILY STRUCTURE

In the future, family structures may undergo a major transformation due to technological developments, social changes and cultural evolution. However, basic human relationships, love, respect and bonds will always be one of the fundamental building blocks of societies. Here are some predictions about how future family structures and relationships may take shape:

Evolution of family structure

In the future, family structures may become more flexible and diverse. As social interactions change under the influence of technology, different family models may emerge:

• **Cohabitation models:** Traditional families may exist in different forms in the future. In addition to nuclear families, different family structures may become more common due to social and economic conditions, where long-term friend groups and multi-generational families live together. This may allow family members to become more dependent on each other or to live more independently.

• **Flexibility of family and kinship ties:** The communication opportunities provided by technology can strengthen ties between relatives by eliminating geographical distances. In virtual environments, family members and relatives can communicate frequently and maintain emotional ties independent of physical distance.

• **Multi-generational living:** Elderly people in families may have longer life spans with the development of genetic engineering and biotechnology. This may lead to multi-generational living models becoming more common. In other words, parents, children and grandparents may live together in large families.

Marriages and relationships

Marriages may be further shaped by social, cultural and economic factors in the future.

• **Diversification of marriage models:** The institution of marriage is likely to take on more flexible and diverse forms in the future. Different models such as open marriages, polygamy, and social partnerships may become increasingly common. People can shape their social ties not only within legal and religious frameworks, but also according to their own needs.

• **Marriage with technological assistants:** New dimensions in relationships with artificial intelligence and biotechnology may make it easier for people to meet suitable partners. Artificial intelligence-supported matching systems may help relationships to be more sustainable and harmonious. In addition, elements such as emotional intelligence and attachment theories may become important in marriages.

• **Deepening of emotional connections:** In the future, the bonds between husbands and wives may deepen by using virtual reality and artificial intelligence. People can get closer to each other through virtual environments, which may increase the emotional intensity of relationships.

Children and family ties

Children's education, development and place in the family may differ under the influence of technology.

• **Parent-child relationship:** Technology can enable parents to educate their children more consciously. Advanced educational platforms and AI-supported applications can enable parents to follow their children's development more closely and provide them with more individualized educational experiences. This can support deeper love and bonding in parent-child relationships.

• **Genetic and biotechnological changes:** Children's health and intelligence levels can be improved through genetic engineering. This can change parents' expectations of their children and enable children to be more biologically durable or long-lived. Children's development can be better monitored with biotechnological support.

• **Children's psychological needs:** With increasing technological developments, children's psychological and emotional needs will become more important. AI-supported systems can analyze children's moods and guide parents. However, excessive use of technology can lead to feelings of loneliness or disconnection among children. Therefore, it will be important for families to maintain strong bonds with their children.

Love and respect within the family

The understanding of love and respect within the family will be shaped according to the values and cultures of societies. However, there may be some important changes:

• **Digital and emotional ties:** In the future, family members will be able to be closer to each other in a virtual environment. Families will be able to understand and meet each other's emotional needs more easily with technologies such as video calls, shared virtual activities, and augmented reality (AR).

• **Roles in the family:** Traditional roles between women and men may change over time, and an egalitarian structure may come to the fore. Women's participation in the workforce and men sharing responsibilities within the home may become more common. This may lead to healthier and more respectful relationships between spouses.

• **Empathy and emotional intelligence:** Artificial intelligence and psychological analysis tools can improve the emotional intelligence of family members. People can become more empathetic, better understand each other, and establish healthier relationships.

Relationships between husband and wife

Husband-wife relationships may develop a different dynamic with social norms and cultural changes.

• **New methods of communication:** In marriages, the way couples communicate may change. Relationships between spouses can be managed more healthily with artificial intelligence-based therapies or digital family counseling. This can allow couples to communicate more openly and honestly.

• **Technological support in marriage:** Couples can better understand each other's needs with artificial intelligence-supported systems. For example, couples who become compatible with smart home systems in sharing routine tasks at home can live more productive and less stressful lives.

• **Deepening emotional bonds:** In the future, it is expected that artificial intelligence and virtual environments will strengthen the emotional bonds in couples' relationships. Thanks to virtual experiences, couples can spend time together in ways they could not have imagined before and discover different dimensions in their relationships.

In summary, family structures and relationships may become more flexible, dynamic, and technologically advanced in the future. However, no matter how much technology develops, love, respect, bonds, and human values will still be the most important elements within the family. Families can use technology to establish healthier relationships, strengthen emotional bonds, and maintain a sharing lifestyle. Relationships within the family will become deeper, more understanding, and more participatory over time.

THE FUTURE OF RELIGIONS

The future of religions will be shaped by social, cultural and technological developments. However, in a historical context, religions tend to transform rather than disappear completely as long as humanity exists. There are different scenarios and expert opinions on this subject:

Is it possible for religions to disappear?

• **Its role in human history:** Religion is not just a belief system for people, but also a structure that provides social order, determines moral rules, and responds to the search for meaning.

• **The search for meaning:** No matter how much technology develops, basic human issues such as death, the purpose of existence, and ethical questions will continue to exist. This makes it difficult for religions to disappear completely.

• **Modern secularization:** Although the influence of religion on social life has diminished in some regions, as seen in Western societies, religious beliefs are still strong worldwide.

Transformation of religions

• **New religions and belief systems:** In the future, new spiritual movements and philosophies blended with technology, artificial intelligence and science may emerge.

• **Digital worship:** Religious rituals can be performed via virtual reality (VR) platforms.

- **Individual beliefs:** More individual and flexible belief systems can develop by moving away from institutional religions.

Relationship between science and religion

- **Scientific developments:** The explanation of some mysteries in the universe by science may question the role of religions. However, this situation may reshape spiritual quests rather than completely eliminating religion.

- **Science-religion balance:** Rather than conflicting with each other, religion and science may continue to exist in harmony in some societies.

Cultural and social factors

- **Social solidarity:** Especially in times of crisis, religions can continue to be an important tool to strengthen social ties.

- **Migration and cultural change:** With globalization, interaction between different religions will increase.

AI and religion

- **AI religions:** Some futurists predict that AI could develop its own ethical systems, creating a kind of "digital religion."

- **Artificial spirituality:** A future in which people seek personal spiritual guidance through AI is possible.

In conclusion, religions are unlikely to disappear completely, but it is clear that they will transform and continue to exist in different forms. As people seek answers to their existential questions, belief systems will continue to be a part of life in some way, even if they are digitalized.

TRADITIONAL AND INNOVATIVE SPORTS

The future looks set to see a balance between physical sports and e-sports. While technological advances, social trends and factors such as genetic engineering will influence the evolution of sports, it is unlikely that physical sports will disappear completely. Instead, the form, rules and experience of sports may change, but human interaction and physical abilities will still play a significant role. Here are some predictions for how sports could evolve in the future:

1. The future of physical sports

Physical sports are ancient traditional activities in which people test their bodies and mental endurance. Such sports will continue to play an important role in the cultural heritage and social ties of societies.

Evolved physical sports

• **Technological integration:** Physical sports will be integrated with technological developments in the future. For example, traditional sports such as football, basketball, and tennis will use wearable technologies and sensors to monitor and improve player performance. These devices can help players improve their skills by providing instant feedback.

• **Hybrid sports:** With the combination of e-sports and physical sports, "hybrid sports" may emerge. These sports combine physical movement with virtual interaction. For example,

AR (augmented reality) or VR (virtual reality) technologies can create new games where players are physically active on the field but interact with virtual elements.

• **Universal popularity:** Despite cultural diversity, interest in sports such as football, basketball, and tennis will continue to be largely worldwide. However, sports can evolve by different cultures in a globalizing world and local variations may emerge. For example, a sport that is very popular in one country may be less common in another, but will still remain a means of global cultural connection.

The impact of physical sports on health and education

• **Health benefits:** Physical sports will continue to be an important tool for health and fitness. In particular, the positive effects on mental health will ensure that sports continue to play a central role in people's lifestyles.

• **Physical sports in schools and education:** The importance of sports that support children's physical development will increase in education systems. Physical education and sports activities will continue to be included in school curricula in the future, because physical health and skills such as teamwork will continue to be important.

2. The rise of e-sports

The rise of e-sports will be a trend that develops simultaneously with physical sports. However, e-sports and physical sports will not exist as rivals, but as two areas that evolve in parallel.

The relationship between e-sports and physical sports

• **Combining physical and mental performance:** E-sports allow players to compete in an environment that requires strategy and mental stamina. However, physical sports are not at risk of disappearing completely because they offer a full experience that benefits from the combination of body and mind.

• **Virtual sports arenas:** E-sports will create global arenas similar to traditional sports. With the merging of the arts and gaming industries, sports events will also be held in the virtual world and will appeal to a very large audience.

• **E-sports and sports:** E-sports players will train and develop mental stamina like traditional athletes. However, physical exercise programs will also be required to offset the effects of e-sports on physical health.

3. Evolution of the human body and sports

• **Genetic and physical abilities:** In the future, with genetic engineering and biotechnology, people will be able to make their physical abilities more efficient. This will push athletes' physical limits even further. However, such developments are likely to create ethical problems in sports. For example, new rules may be created to ensure that genetically superior athletes compete fairly.

• **The combination of mental abilities and the body:** While people's mental abilities are being developed with virtual intelligence and artificial intelligence, physical sports may also diversify with new formats based on brain-body interaction.

4. Digitalization of sports and new spectator experiences

• **Advanced viewing experiences:** Art, media and technology will merge to create new types of viewing experiences. Viewers can watch matches in a virtual world, immersed in 360-degree video, virtual reality (VR) and augmented reality (AR) technologies that are specific to their own experiences.

• **E-sports and physical sports interaction:** In the future, physical sports and e-sports could merge, integrating digital simulations and physical games. For example, in sports such as football, virtual competition areas could be created that present viewers with a virtual version of the moves the players make.

Artificial intelligence and technological advances will change the nature of sports and enable the emergence of new sports. The integration of technology into sports will create entirely new types of events, combining both the physical and digital worlds. Here are some innovative sports that could emerge in the future:

1. AI-powered sports

Artificial intelligence-supported sports could bring a new understanding of sports where players' physical skills are combined with artificial intelligence, strategy and guidance.

• **AI wars:** Humans team up with AIs to play strategic games. These games combine human skills with AI's strategic intelligence, resulting in real-time intelligence wars.

• **AI football or basketball:** Human players use AI algorithms' suggestions to make more effective and strategic moves. These sports will be based on physical skills and strategy, as well as teamwork.

2. Virtual reality (VR) sports

Advanced VR technology will enable players to experience sports from virtual to real, creating new games where the physical and digital worlds merge.

• **Virtual football/basketball:** With VR glasses, players will be virtually in a stadium and will move on the field and compete with virtual players. VR technology will bring all the excitement of real-world sports to life in a virtual environment.

• **Virtual races and simulations:** With VR technology, people can participate in virtual races without straining their physical conditions. This allows them to compete in activities such as flying, racing cars or water sports as if they were real-world sports.

3. Hybrid sports (physical and digital integration)

Sports that will emerge from the merging of the physical and digital worlds will require both physical skills and digital intelligence.

• **Augmented reality (AR) sports:** Virtual elements can be interacted with using augmented reality technology in a physical playing field. For example, virtual opponents or targets can be added to the football field, allowing players to compete both physically and virtually.

• **Football 2.0:** With the combination of AI and AR, players can play a football match with virtual obstacles or hidden targets. For example, players can develop strategies to overcome virtual obstacles as well as pass the ball.

1. Biotechnological sports

Advanced biotechnology could enable entirely new sports to emerge by improving human bodies.

• **Genetic sports:** Genetic engineering could make new types of sports possible by improving athletes' performance. For example, humans with increased endurance, enhanced speed, or special strength abilities could demonstrate their physical abilities in new sports.

• **Biotechnological aptitude tests:** Sports may emerge where people use biotech implants and advanced sensors to track their performance in real time. These sports will push physical boundaries and explore how far the body can go.

2. Neuro-interactive sports

The use of neural connections in sports may lead to a new branch of sports that consists of connecting the brain directly to a virtual system.

• **Neuro-sports:** Sports managed by brain waves will test not only the physical but also the mental capacities of the players. For example, players can create strategies with their brain power while performing physical movements and play games with their minds, not just their bodies.

• **Mental competitions:** Sports controlled by brain waves might involve mental races against digital targets. For example, a player might compete against opponents in virtual races using mental concentration and quick thinking.

3. Mind and body experience sports

The impact of technology on the human body could enable the mind-body connection to create new sports.

- **Synethesis of mind and body:** In the future, the boundaries between human mental abilities and physical performance may become even more blurred. With artificial intelligence or neural networks, the interaction between the brain and the body can be accelerated, and athletes can direct their physical movements with brain commands.

- **Physical and sensory sports:** New sports branches that take place in the virtual world and require not only physical but also sensory and mental participation of the players may develop. For example, players may compete in a completely sensory way, not physically, in special areas created by art, music or lights.

4. Holographic sports

Players who compete with each other in real time using holographic projections will be able to compete not only in the physical field but also in the virtual world.

- **Holographic boxing or fighting:** Virtual fighters created with holographic projections can compete. These fighters take place in the arena as realistic 3D holograms without physical combat and compete against real athletes.

- **Holographic football or basketball:** These sports can become holographic events that combine artificial intelligence players with real players in the same environment, creating a physical competition in a virtual environment.

In summary; physical sports will definitely continue to exist in the future, but technological advances will change the way sports are experienced. E-sports will become increasingly popular, but the social and cultural role of physical sports will still be strong. While people may be interested in traditional sports that improve physical health and mental resilience, they may also prefer to compete in the virtual world. As a result, physical sports and e-sports will develop in parallel and continue to exist in a way that complements each other. Technologies such as artificial intelligence, biotechnology, art and virtual reality will make sports not only a physical experience, but also a digital, mental and sensory experience. As the boundaries between physical sports and digital sports become increasingly blurred, people will participate in various competitions and competitions in both the real world and the virtual environment. These innovative sports will offer new opportunities in terms of both entertainment and health.

VII

THE FUTURE OF HEALTH AND NUTRITION

HOSPITALS AND HEALTH SERVICES

The future of healthcare will undergo a major transformation with advances in technology, biotechnology, artificial intelligence, genetic engineering, and personalized medicine. Healthcare areas such as hospitals, surgeries, cancer treatments, heart diseases, brain surgeries, and drug treatments will transform into systems that offer much faster, more efficient, and personalized solutions.

Hospitals and health infrastructure

• Smart hospitals and digitalization: In the future, hospitals will be smart hospitals equipped with artificial intelligence and IoT (internet of things) technologies. These hospitals will automatically collect and analyze patient data and optimize treatment processes. Digital health records will provide instant access to patients' past health conditions.

• Autonomous robots and assistive technologies: Autonomous robots will play an important role in surgeries, treatment applications and patient care management. For example, robotic surgeries will enable more precise and minimally invasive procedures. In addition, robots will also take part in patient transportation, drug distribution and care.

• Telemedicine and remote healthcare: Telemedicine and virtual clinics will eliminate the distance between the patient and the doctor, making remote examinations, treatment recommendations and follow-ups possible.

This will provide great convenience, especially for patients in remote areas or with limited mobility.

• Artificial intelligence-supported diagnosis and treatment: Artificial intelligence will be the most important aid in early diagnosis of diseases, risk analysis and treatment processes. Artificial intelligence will examine millions of patient data and offer doctors the correct diagnosis and treatment options.

Surgeries and emergency interventions

• Minimally invasive surgery: In the future, minimally invasive surgical techniques that require less recovery time will become widespread. Thanks to robotic surgical systems, surgeons will be able to perform much more precise and controllable operations. Major surgeries will become less risky with laparoscopic and endoscopic methods.

• Real-time data usage: Real-time data flow will be provided during surgeries. Doctors will be able to monitor information such as the patient's biometric data, blood sugar, heart rate, and oxygen levels in real time. This will increase the success rate of surgeries.

• Autonomous surgeons: Robotic systems supported by artificial intelligence will be able to perform certain surgeries autonomously. By preventing errors made by human surgeons, surgeries will be performed faster and more accurately.

• Artificial intelligence in emergency interventions: In emergency services, ARTIFICIAL INTELLIGENCE systems will be able to quickly analyze patient conditions, determine which patients require more urgent intervention, and help doctors make the right decisions.

Cancer treatments

• Personalized cancer treatment: Thanks to genetic tests and biomarkers, personalized cancer treatment plans will be created for each individual. This will make it possible to diagnose cancer at earlier stages and to direct treatment according to personal genetic structure.

• Immunotherapy and genetic engineering: Immunotherapies will make the immune system more effective against cancer cells. In addition, cancer cells can be directly intervened with genetic engineering. CRISPR technology will change the genetic structure of cancer cells and prevent the spread of cancer.

• Targeted treatment with nanotechnology: Nanotechnology will also revolutionize cancer treatment. Nanoparticles will be directed directly to cancer cells and will be able to treat healthy cells without harming them.

• Digital cancer monitoring: Cancer patients will be able to follow their treatment processes via digital platforms. The progress of cancer will be monitored with real-time biomarker analyses and artificial intelligence algorithms.

Hearth diseases and brain surgeries

• Personal heart health monitoring: Thanks to wearable technologies, individuals will be able to continuously monitor their heart health. Digital stethoscopes and heart rhythm monitors will help detect heart diseases in advance.

• Artificial intelligence-supported cardiology: In heart diseases, artificial intelligence algorithms will support the decisions of cardiologists. Narrowing of the heart vessels will be detected much more precisely with ARTIFICIAL INTELLIGENCE-based imaging systems.

• Artificial heart and organ transplantation: Artificial heart technologies will replace the natural heart functions of humans. In addition, organ transplantation processes will become more efficient with bioengineering and 3D printing technologies.

• Brain surgeries and neurological treatment: Artificial intelligence and robotic surgeries will play a major role in brain surgery. For neurological diseases, devices that monitor brain activity will be able to predict diseases in advance. Thanks to brain-computer interfaces (BCI), paralyzed patients will be able to communicate directly with the devices.

• Neurotherapy and genetic interventions: Genetic therapy

and neurological treatment techniques will be developed for brain diseases. Genetic engineering will be effective in the treatment of diseases such as Parkinson's and Alzheimer's.

Drug therapy and new treatment methods

• Genetic and personalized medicines: Thanks to pharmacogenetics and genetic tests, drug treatments specific to each individual will be developed. Drugs will be optimized according to personal genetic structure, thus providing more effective and less side effect treatments.

• Biotechnological drugs and vaccines: New biotechnological drugs will offer more specific and targeted treatment methods. Treatment processes will be more efficient thanks to targeted biological drugs and genetic engineering.

• Targeting drugs with nanotechnology: Nanotechnology will enable drugs to be directed directly to disease cells. As in cancer treatment, drugs will be carried to the right target in the body thanks to nanobots.

Healthcare of the future

In the future, healthcare services will be faster, more sensitive and more accessible thanks to artificial intelligence, biotechnology, nanotechnology and personalized treatment methods. While digitalization and remote healthcare services will enable people to live healthier lives, robotic surgeries and genetic treatments will revolutionize the treatment of major diseases. Personalized healthcare solutions will enable early diagnosis of diseases and treatment processes will become more efficient.

HAIR IMPLANT AND STEM CELL

Stem cell treatments

Hair follicle renewal: It will be possible to regenerate hair follicles using stem cells. Thanks to this method, even completely lost hair follicles can be revived.

Personalized treatment: Solutions developed from the person's own stem cells will eliminate the risk of rejection by the body and provide more natural results.

Gene therapy

Genetic manipulation: As the genetic causes of hair loss are resolved, it will be possible to eliminate the genetic factors that cause hair loss with gene treatments. Permanent solutions: Thanks to genetic corrections, hair loss can be prevented for life with a single treatment.

Artificial hair technologies (bio-materials

Nano hair implants: Hair transplantation surgeries can become more practical thanks to implants developed using biomaterials that are very similar to natural hair and compatible with the body.

Smart hair implants: Biocompatible artificial hair implants that can simulate hair growth and maintenance over time can be developed.

Tissue engineering

Scalp laboratory production: Permanent solutions can be provided by transplanting scalp and hair follicles produced in a laboratory environment to areas with hair loss.

Full hair reconstruction: Thanks to tissue engineering, it will be possible to completely reconstruct natural hair growth.

Pharmaceutical and serum technologies

Molecular treatments: A new generation of molecular drugs that activate hair follicles and stimulate dormant follicles may be available. Hair stimulating serums: Advanced serums and sprays will not only prevent hair loss but also accelerate new hair growth.

Bio-printing (3D bioprinting)

Follicle printing: A method of "printing" hair follicles directly onto the scalp using 3D bioprinters may be developed. This technique could enable fast and low-cost hair transplants.

Smart cosmetic technologies

Laser and LED therapy devices: Laser and LED beams that stimulate hair growth may be used at home in more advanced forms in the future. Artificial intelligence-supported treatments: Artificial intelligence applications that analyze individual causes of hair loss and offer personalized care recommendations may become widespread.

Hormonal balancing treatments

Hormone therapies: Hormone-based treatments will become more effective, especially against hair loss problems caused by testosterone imbalance.

In the future, hair loss will cease to be just an aesthetic problem and will become a condition that can be solved radically. Thanks to stem cell treatments, genetic corrections and biotechnological developments, people will be able to have hair of the density and natural appearance they desire. In the future, housework will transform in a way that will make people's lives easier and save time thanks to developing technologies. With artificial intelligence, robotic systems, automation and smart devices, many tasks at home will become almost completely automated. Here are the predictions:

CAPSULE FOODS

Developments in food technology and population growth may radically change our eating habits in the future. Capsule foods stand out as practical and nutrient-dense solutions.

Features of capsule foods and drinks

1.Dense nutritional content: One capsule can contain daily needed vitamins, minerals, proteins and fats.

2. Easy consumption: It offers fast consumption without wasting time. It can be preferred especially in busy city life.

3. Long shelf life: It is resistant to deterioration and can be stored for a long time.

4. Personalized nutrition: Personalized capsule diets can be designed based on genetic structure or daily activities.

Areas of use of capsule foods

1. Space travel: Condensed capsule-like foods are already used for astronauts.

2. Intense city life: They can be quick nutrition alternatives for office workers or individuals with time constraints.

3. Military and disaster situations: In disaster areas or war conditions, capsule foods can be used as a survival tool.

4. Sports and performance foods: Energy and protein capsules can provide more effective nutritional solutions for athletes.

5. Medical nutrition: Customized medical nutrients in capsule format can be used for patients with swallowing difficulties.

Transforming food in the future

• **Foods produced with 3D printers:** Thanks to food printers, meals customized to individual tastes can be printed.

• **Synthetic meat and plant-based foods:** Meats produced in laboratory environments and more sustainable plant-based protein sources will become widespread.

• **Nanotechnological foods:** Smart food capsules that slowly release into the body and track health parameters.

• **Taste and smell simulations:** With virtual reality-supported systems, the brain can experience different tastes and smells while consuming capsules.

Advantages

• Fast and practical consumption.

• Adaptable to personal health conditions.

• Sustainable and environmentally friendly production.

• Long-lasting products that are resistant to deterioration.

Disadvantages

• Loss of traditional food culture.

• Decrease in emotional and social eating experience.

• Risk of addiction to chemical ingredients.

• Deprivation of taste and texture.

In summary; capsule foods may gain an important place in our lives as a result of intense living conditions and the need for sustainability. However, future eating habits should be balanced not only with practicality but also with emotional and cultural experiences. A balance between traditional food culture and high technology seems inevitable.

SOILLESS FARMING

Future potential of soilless agriculture

Resource efficiency

• Soilless agriculture saves up to 90 percent of water compared to traditional agriculture.

• It offers environmentally friendly production with less fertilizer and pesticide use.

Production in cities with vertical farming

• Vertical farming applications make farming possible even in big cities.

• Reduces transportation costs and carbon footprint.

Production independent from climate

• Production done in closed environments is not affected by weather conditions.

• It offers the opportunity for uninterrupted agriculture in all four seasons.

Technology integration

• Precision agriculture can be implemented with artificial intelligence, IoT and automation systems.

• The amount of nutrients and water that plants need can be optimized with sensors.

Alternative protein sources

• Microgreens, algae and other plant protein sources can be produced more efficiently with soilless agriculture.

Space farming

1. NASA and private companies are testing soilless farming methods for space colonies.

2. In the future, food production with these techniques may be possible on Mars and the Moon.

Challenges to be encountered

• **High initial cost:** Setup costs are still high compared to traditional farming.

• **Energy dependency:** Energy consumption is a significant factor in closed systems.

• **Requirement for technological expertise:** Requires more technical knowledge than traditional farming.

Soilless farming methods

1. Hydroponics (water culture)

Plants are grown by immersing their roots directly into water enriched with nutrients. This system provides all the minerals the plants need in solution.

Advantages

• Saves water (90 percent less water than traditional agriculture).

• Fast growth and high yield.

• No risk of harmful weeds and soil diseases.

2. Aeroponics (air farming)

Plant roots are suspended in the air and nutrient solutions are applied to the roots by spraying.

Advantages

• Minimum water consumption.

- Faster root development due to high oxygen levels.

- Clean and hygienic environment.

3. Aquaponics (integration of fish and plants

In this system, where fish farming and agriculture are combined, fish waste is transformed into a natural food source for plants.

Advantages

- Creates a cyclical ecosystem.

- Fish and plant production can be done simultaneously.

4. Substrate culture (solid medium cultivation)

Instead of soil, rock wool, perlite, coconut fiber, etc. are used. These media ensure the transmission of water and nutrients to the plant roots.

Advantages of soilless agriculture

- **Less water use:** 70-90% less water consumption than traditional agriculture.

- **High yield:** When optimum conditions are provided, higher yields can be obtained compared to traditional agriculture.

- **Environmentally friendly:** No need for pesticides and harmful chemicals.

- **Climate independence: Production can be done in closed systems in all seasons.**

- **Vertical farming applications:** More product production becomes possible with less space in cities.

Soilless agriculture in the future

Smart agriculture and automation: Agricultural processes will be optimized with sensors, artificial intelligence and robots.

Urban agriculture: Food production will be possible in metropolitan areas with skyscraper farms and vertical farming applications.

Solution to the climate crisis: It will become an important tool in combating drought and desertification problems.

Mars and space agriculture: Organizations such as NASA are researching soilless agriculture techniques on Mars and space stations, and it is not a dream to do agriculture there in the future.

In general, soilless agriculture seems to be one of the cornerstones of sustainable food production in the future. How widespread do you think this method can be in Turkey?

Soilless agriculture is a modern agricultural technique in which plants are grown in nutrient solutions, water or various solid media instead of soil. This method is seen as the key to sustainable agriculture in a world where agricultural areas are shrinking, water resources are becoming scarce and the effects of climate change are increasing. Soilless agriculture is seen as one of the key solutions of the future in terms of sustainable food production and environmentally friendly agricultural techniques.

REST AND SLEEPING PATTERNS

In the future, people's sleep patterns will likely change significantly due to advances in technology, health, and lifestyle. Here are some possible scenarios:

• **Personalized sleep management:** Thanks to wearable devices and biometric sensors, sleep patterns will be optimized according to personal needs. People will be able to determine their bodies' most efficient sleep times and get better quality rest in a shorter time.

• **Shortening sleep durations:** Thanks to neurological research and biotechnology, the repair processes the brain performs during sleep can be accelerated. This can allow people to get by with 4-5 hours of sleep instead of 8 hours.

• **Sleeping pills and neurotechnology:** Drugs or neurotechnological solutions (such as brain stimulation) can be developed that can replace sleep. These solutions can allow people to stay alert throughout the day.

• **The importance of dark and light cycles may decrease:** Factors such as space colonization or living in closed areas with no day-night cycle may require biological clocks to be managed with artificial light.

• **Social and cultural changes:** Intense work tempos and 24-hour economies may change the meaning of day and night. The "polyphasic sleep" (several short naps per day) model may become more common.

With these changes, sleep seems to be a part of not only resting but also productivity and health management.

In the future, the way people rest may change significantly with the advancement of technology, neuroscience, biotechnology and changes in social structure. Here are possible resting methods:

• **Rapid neurological renewal (brain rest devices):** Mental fatigue can be relieved with just a few minutes of application thanks to brain stimulation devices. These devices can provide rapid renewal by artificially stimulating the areas of the brain that are activated during rest.

• **Virtual reality (VR) rest areas:** People can go on virtual nature walks to escape from stressful and tiring environments, do "virtual" meditation by the ocean, or relax in their favorite vacation spots.

• **Micro rest (power nap) cabins:** With sleep and rest capsules located in offices or public areas, short 10-15-minute sleep sessions can become standard.

• **Artificial intelligence-supported meditation and breathing therapies:** People can manage their stress more quickly thanks to meditation sessions and breathing techniques personally adapted with artificial intelligence.

• **Biotechnological supports (anti-fatigue pills):** Biological support pills or injections that eliminate fatigue and accelerate cell renewal can be developed.

• **Sensory isolation tanks:** Isolation tanks filled with salt water can provide both mental and physical relaxation by making the body feel completely free of gravity.

• **Time-compression sleep technologies:** People can feel like they have slept for hours with a few minutes of "deep sleep" simulations.

• **Areas integrated with nature:** Increasing green areas and living spaces integrated with nature can encourage people to rest faster with natural methods.

With these methods, the concept of rest can go beyond traditional long-term sleep and vacation and become more efficient and technological.

VIII

TRANSPORTATION, TRAVEL AND SAFETY IN THE FUTURE

TRANSPORTATION FACILITY

In the future, intercity and international transportation will undergo a major transformation. Thanks to high technology, sustainable energy systems and automation, transportation will become faster, more efficient, more environmentally friendly and more accessible. Here are some predictions about how the transportation system will evolve in the future:

Autonomous vehicles and rapit transit systems

- **Autonomous vehicles:** Driverless cars, buses and trucks will make transportation safer and more efficient. These vehicles will also be used for intercity and international travel. Artificial intelligence will determine routes according to traffic density, shortening travel times and minimizing accidents.

- **Hyperloop:** High-speed vacuum tube transportation systems such as the Hyperloop will revolutionize intercity transportation. This technology will make it possible to travel much faster than the surface of the earth. People will be able to travel from one city to another in a very short time. For example, it will be possible to reach Ankara from Istanbul in 30 minutes.

- **High speed trains:** Electric and magnetic levitation (maglev)-based high-speed trains will make road travel faster and more efficient. They will play an important role in intercity transportation and reduce air pollution.

Developments in the air transportation

- **Vertical take-off and landing (VTOL) vehicles:** Flying cars or vertical take-off and landing drones (VTOL) will be used especially for short-distance transportation within and between cities. These vehicles will provide faster and more efficient transportation by avoiding congested roads. It will also eliminate the time loss to get from city centers to airports.

- **Electric airplanes:** Electric planes and zero-emission flight technologies will make air travel more environmentally friendly. Electric planes will become more common for national and international travel, offering a cheaper and more environmentally friendly alternative for short distances.

- **Autonomous planes:** In the future, autonomous aircraft will travel instead of piloted planes. These planes will become safer with artificial intelligence and advanced navigation systems, increasing passenger safety.

Sustainable energy and environmentaly friendly transportation

- **Electric vehicles:** Electric vehicles will make urban and intercity transportation more environmentally friendly. Electric vehicles and charging infrastructures will be much more advanced and will help decarbonize transportation. Electric buses and freight transport will also benefit from this transformation.

- **Solar powerd vehicles:** Especially in long-distance travel, solar-powered vehicles and buses will come into play. This will protect the environment by consuming less energy and using renewable energy sources.

- **Zero emission transportation solutions:** Zero-emission transportation In the future, zero-emission transportation solutions will cover all transportation networks. The use of fossil fuels in both land, sea and air transportation will gradually decrease and be replaced by clean energy sources such as hydrogen, electricity and solar energy.

Drones and autonomous transportation

• **Cargo and freight transportation:** Drones will revolutionize freight transportation, as well as intercity transportation. Cargo drones, especially those carrying goods to remote areas, will offer fast, cheap and safe transportation solutions.

• **Autonomous freight vehicles:** Vehicles such as trucks and lorries will be equipped with autonomous technologies and will travel automatically, reducing traffic congestion. These vehicles will increase the speed of freight transportation.

Smart transportation infrastructures

• **5G and smart transportation systems:** Thanks to 5G technology, vehicles will be able to receive real-time data and constantly update traffic conditions, weather conditions and routes. This will enable vehicles to move more efficiently and prevent traffic congestion.

• **Intelligent traffic management:** Urban and intercity traffic will be controlled by artificial intelligence-supported traffic management systems. These systems will optimize traffic flow, ensure coordination between vehicles, and reduce the risk of accidents.

Hybrid travel modes and flexibility

• **Integrated travel solutions:** In the future, integration will be achieved between different modes of transportation. For example, a person will travel by public transportation within the city and use high-speed trains for intercity distances to reach the destination in the fastest way. All modes of transportation will be usable together with digital ticketing systems and mobile applications.

• **Capsule transportation and mobile peace points:** People will be able to spend time comfortably during their journey while traveling with autonomous transport vehicles. Long journeys will become more comfortable by offering services such as mobile offices, sleeping capsules or rest areas.

Space transportation

• **Space travels:** In the long term, space transportation may also be on the agenda. Space tourism and space travel, which are currently only attempted by private companies, will become more accessible. In particular, transportation to extraterrestrial colonies can be carried out by overcoming large distances in space.

Future of transportation

In the future, transportation will be reshaped by automation, sustainable energy, artificial intelligence and innovative technologies. People will have faster, safer and more environmentally friendly transportation systems. Increasing integration in transportation will enable more efficient use of different transportation modes and transportation will be provided at lower costs. People will be able to travel faster and make the world more accessible with transportation solutions that reduce time and energy loss.

VISA FACILITY FOR FOREIGN TRIPS

There may be significant changes in the future in issues such as international travel, citizenship, digital citizenship, visa, passport, immigration and reactionism. These changes will be shaped by factors such as technological developments, international relations, global security and social dynamics. Here are some possible developments in these areas:

International travel and transportation

With the advancement of technology, travel can become faster, safer and more efficient:

- **Fast transportation vehicles:** Advanced means of transportation, such as supersonic aircraft, hyperloop systems, or space travel, could transport people from one place to another in seconds. This could make international travel much easier and largely eliminate the barriers of time and distance.

- **Virtual trips:** echnologies such as virtual reality (VR) and augmented reality (AR) can eliminate the need for physical travel. People can travel in virtual environments and experience other cultures, cities, and natural areas.

Digital citizenship

In the future, the concept of digital citizenship may become more important. With technological and social developments, the physical borders of countries and the definition of citizenship may change:

- **Digital identity and citizenship:** With blockchain technology and artificial intelligence, people can have digital identities. These digital identities can facilitate people's access to basic rights such as social security, healthcare, and education in different parts of the world. Digital citizenship can provide access to services without the need for physical citizenship.

- **Multiple citizenship:** In the future, citizenship of more than one country may be acquired through digital platforms. Systems that provide digital residence or foreign digital citizenship may make citizenship relations between countries more flexible and multi-layered.

Visa and passport

In the future, visa and passport processes may experience significant changes, especially with digitalization:

- **Automatic travel permits:** Thanks to AI and biometric scanning technologies, travel could become much faster. Visa applications could be automated, and some countries could even digitally identify travelers in advance, shortening or eliminating visa processes altogether.

- **Digital passports:** Thanks to AI and biometric scanning technologies, travel could become much faster. Visa applications could become automated, and some countries could replace physical passports with digital documents, such as blockchain-based passports, that can digitally identify travelers, reducing the risk of fraud and making passport control processes more efficient and secure. This could shorten or eliminate visa processes altogether.

Migration and asylum

Migration and asylum will be shaped by different dynamics in the future. The development of technology can affect international mobility and social structures:

- **High mobility:** People can move to different countries more easily thanks to digital work models, remote working and globalization. However, factors such as international travel, economic opportunities and social security can affect migration flows.

- **Climate change and refugees:** Climate change and natural disasters can trigger migration. Climate refugee issues may become an important issue in the coming year, and new international policies and migrant support systems can be developed to address this.

- **Social security agreements:** Social security agreements and refugee admission protocols can be developed between countries. In this way, the rights of immigrants or refugees can be more easily integrated into the social system of each country.

International relations and travel security

- **Border security and technology:** In the future, border security could become more digital. With artificial intelligence and biometric scanning, human mobility could be controlled more efficiently. However, such security measures could also raise ethical questions about privacy and individual freedoms.

- **International communication and travel:** Social media and communication platforms can shape relations between countries. People can communicate with each other more quickly and directly on digital platforms, and thus get to know different cultures more closely. This can transform understandings of national borders.

The future of travel and communication

- **Fast and unlimited communication:** In travel and international relations, innovations such as virtual reality and teleportation technologies can further eliminate physical distances between people. People can come together through digital avatars or holograms.

- **Environmental sustainability:** Travel could become more environmentally friendly in the future, with electric planes, solar-powered vehicles and modes of transport that reduce carbon footprints becoming more prominent.

Legal and political dimensions

- **International laws and digital citizenship:** Digital citizenship can be a legal issue for different countries. When people have a digital citizenship system from a particular country, they may be subject to international regulations. This may require regulations on issues such as the right to work in foreign countries, citizenship rights, and immigration rights.

As a result; in the future, international travel, citizenship and immigration policies may undergo a major transformation under the influence of technology. Applications such as digital citizenship, blockchain-based passports and automatic visa approval systems may make travel and international relations faster and more efficient. In addition, factors such as climate change, social inequalities and global wars may be among the important dynamics that will affect people's decisions to migrate. These developments may enable international relations to become more flexible, connected and technological. In the future, the concept of being a global citizen and a one-world state may become more widespread depending on technological, social, economic and political developments. However, there are some important factors and possibilities regarding how this will happen and whether it will be possible.

BEING A WORLD CITIZEN

Will it become easier to become a world citizen?

In the future, the idea of being a global citizen may become more attainable due to several important factors:

- **Digital citizenship:** Thanks to digitalization and blockchain technologies, people can not only have citizenship of a certain country, but also gain digital citizenship. This can give people access to some legal rights and services around the world, but to become a full-fledged "world citizen", all world governments must accept this system.

- **Globalization and globan methods:** With globalization, economic and trade relations have become more interconnected. This has made people more mobile. Thanks to technological advances, people can work, live and study in other countries. In the future, human mobility around the world may become easier, but the social, economic and cultural policies of each country can still be a major obstacle.

- **Legal and social obstacles:** However, at present, each country has its own legal system and citizenship laws. This poses a significant obstacle to the idea of becoming a "global citizen." Creating a single citizenship law worldwide may be very difficult due to factors such as national sovereignty, the desire of societies to preserve their identities, and local cultures.

Can the world be governed by the single source?

The question of whether the world can be governed by a single center can be answered depending on many different perspectives:

• **One world state (global government):** While this idea has historically been present in many utopian ideas and some political ideologies, in practice its realization faces major social, cultural, and political obstacles. National sovereignty and state independence are among the greatest obstacles to the establishment of a single worldwide government. However, some global challenges, such as climate change, global pandemics, and international security issues, may require greater cooperation and global coordination. Over time, this may lead to a tendency toward strengthening of some global governance structures, or even toward a world government.

• **A single management system:** Today, various international organizations (such as the UN, the World Trade Organization, and the World Health Organization) are concerned with some global issues. The increase in such international platforms may lead to a more centralized understanding of global governance. However, such a monocentric governance will face the need to protect national sovereignty and cultural differences.

• **Technological effects:** Technologies such as artificial intelligence, blockchain, and cybersecurity could enable a more centralized model of global governance. These technologies could facilitate faster data flows, international cooperation, and governance processes. However, this could also raise new ethical issues such as privacy and democratic oversight.

Major societal obstacles and challenges

• **Cultural differences:** The cultural, religious and linguistic diversity around the world poses a major obstacle to the adoption of a single governance structure. Each region has its own social structure, traditions and values, which can make it difficult for a centralized government to function.

- **Political resistance:** Many countries around the world may resist the loss of their sovereignty and the idea of an external administration. National independence and political autonomy are still among the fundamental values of many countries.

- **Economic differences:** There is a huge economic disparity globally. The economic gap between developed and developing countries is a major obstacle to the successful operation of a single government worldwide. Rich countries may be reluctant to adopt the same model of government as poor countries.

Alternative management models

In the future, even if the world is not governed by a single center, different governance models may emerge:

- **Autonomy:** A balance can be struck between global unity and regional autonomy. For example, while some global decisions are being made worldwide, local governments can still continue to solve regional problems.

- **Multinational management:** Global organizations such as the United Nations could gain more power, and a model could emerge that would allow worldwide cooperation but maintain independent governments. In this model, national governments could enter into new international agreements to maintain global order.

In summary; it may become easier to be a global citizen in the future, especially thanks to digitalization and technology. However, establishing a single world state may be a very difficult goal because each country has its own sovereignty and cultural structure. Global cooperation and multinational governance models seem to be a more likely scenario. In this process, technology will play a major role and international relations will evolve to be more cooperative, but still preserve local sovereignty and social diversity.

HUMAN SECURITY

In the future, ensuring people's property and life security will become much more effective with the opportunities provided by rapidly developing technologies. However, there will also be new risks and security problems brought about by innovations such as automation and artificial intelligence. Therefore, security measures will be provided with more advanced, integrated and fair systems. Autonomous robots, artificial intelligence, advanced security infrastructures and legal regulations will be the basic elements of this process.

1. Autonomous robots and human security

• **Autonomous robots should not harm humans:** Autonomous robots will be made safe for humans through very strict controls and regulations. The ethical boundaries and security algorithms of artificial intelligence will focus on robots only performing certain tasks and not harming humans. In addition, the behavior of robots will be constantly monitored with ARTIFICIAL INTELLIGENCE control mechanisms and open-code systems. If a robot accidentally harms a human, the control center will immediately step in and stop the robot.

• **Safety training of robots:** The robots will be equipped with advanced training systems, especially for security and emergency management. These trainings will program the robots within the framework of both ethical rules and safety protocols.

226

- **Advanced security analysis with AI:** Artificial intelligence will identify potential threats in advance by performing risk analysis and security predictions. ARTIFICIAL INTELLIGENCE will continuously analyze the environment in which robots operate and will respond quickly in the event of any threat.

2. Thefts, murders and robberies

- **Advanced cameras and monitoring systems:** Security will be further strengthened with facial recognition technology, motion detection systems and smart cameras. Artificial intelligence-supported security systems will detect abnormal situations before they occur and immediately notify security forces.

- **Autonomous security vehicles:** Autonomous drones and robot security guards will patrol certain areas to ensure security. These vehicles will be able to intervene when they detect an abnormal situation and notify the authorities when necessary.

- **Smart city security:** Smart city infrastructures will be integrated with cameras, sensors and artificial intelligence systems. These systems will enable easier detection of crimes and develop proactive solutions to prevent criminal behavior.

- **Forensic Technologies and data monitoring:** With biometric analysis, genetic traces, criminal databases and digital monitoring, tracking criminals will become much easier. Blockchain technology will be used to record the digital footprints of criminals and detect suspicious transactions.

3. Intervention in fight and violent incidents

- **Preventetive intervention with AI:** ARTIFICIAL INTELLIGENCE will predict potential violent situations and send early warnings to security forces for intervention. For example, violent incidents that start on the streets or in public areas will be monitored instantly and directed to the police.

- **Advanced communication systems:** During violent incidents, people will be able to call for help immediately via emergency buttons or smart devices. In addition, the location and condition of the user will be determined in real time via phones or wearable devices in times of violence and transmitted to aid teams.

- **Autonomous security forces:** In some cities, robotic security guards or AI-powered vehicles will be able to actively patrol streets and respond to violent incidents. These systems will be specifically programmed to minimize the use of force and violent intervention.

4. Legal and ethical regulations

- **Legal regulations on AI and robots:** The use of robots and artificial intelligence will be limited by ethical and legal frameworks. These frameworks will provide a basic structure to ensure that robots do not cause harm, do not violate personal rights, and comply with safety standards. In addition, artificial intelligence ethical codes will be developed.

- **Judicial system and AI:** AI can be used to detect criminals faster and make criminal proceedings more efficient. However, when it comes to legal liability, such AI systems will still be under the control of regulatory and judicial bodies.

- **AI, security and human rights:** In terms of human rights, artificial intelligence will be checked to ensure that it respects privacy rights and the principle of justice.

THEFT

Theft has never been considered an "occupation" as an illegal and unethical act. However, the form and motivations of such crimes may change in the future with technological developments. Here are some predictions about the state of theft in the future:

1. Reduction in physical theft

• **Digitalization:** Replacing cash with digital payment systems could reduce physical robberies.

• **Smart home technologies:** The widespread use of security systems (facial recognition, artificial intelligence-supported alarm systems) will make home and workplace burglaries more difficult.

2. Cyber theft and digital crimes

• **Cyber attacks:** In the future, theft of data, cryptocurrencies and digital assets may become more common than physical theft.

• **AI-supported attacks:** Phishing, ransomware attacks, and bank account theft attempts may increase with advanced software.

3. Crive prevention technologies

• **Blockchain:** Storing data securely and transparently will make cyber theft more difficult.

- **AI and big data:** Security forces may have the capacity to prevent crimes by predicting them before they occur.

Legal and social factors

- **More severe sanctions:** Penalties and tracking systems for technological crimes will become more effective.

- **Economic prosperity:** As income inequality decreases, motivations for some crimes may disappear

While physical theft has decreased significantly, digital crimes and cyber theft may pose a greater threat in the future. However, advanced security systems and legal measures will play an important role in preventing crimes.

IX

MEDIA, COMMUNICATIONS AND CULTURAL CHANGES

EDUCATION SYSTEM AND LEARNING

With technology, social changes and digitalization, the education system will undergo a significant transformation, especially in terms of face-to-face education, diplomas, schools, universities and books. With innovations such as distance education, artificial intelligence, virtual reality (VR) and augmented reality (AR), education will become more accessible, personalized and interactive. However, this change will bring a period where different models coexist, without leading to extreme scenarios such as school closures or the complete digitalization of the education system. Here are some predictions about the future education system:

1. Face-to-face education and schools

• **Hybrid education models:** Face-to-face education will continue as a hybrid model, but supported by digital tools. Students will take some courses physically at school, while taking other courses on online platforms or in virtual classrooms.

• **Learning spaces and flexibility:** Schools will not only be places of education, but also areas of experience. Educational institutions will offer social interaction areas for individual development, collaboration and creative projects outside of classes.

• **Evolution of school structures:** School buildings will be shaped by more dynamic structures such as digital classrooms and content production workshops supported by artificial intelligence rather than traditional classrooms. Schools will become social learning centers where physical and digital environments merge.

2. Digital education and distance learning

• **Fully integrated educational technologies:** Distance education will cease to be a temporary solution specific to the pandemic and will become a permanent education model. Artificial intelligence-supported learning systems will monitor students' progress and offer them personalized learning paths.

• **Educational content and access:** Instructors will dynamically create educational content based on students' needs in various fields such as art, engineering, science, etc. Online classes will offer not only video lectures but also virtual interactions, instant feedback, and independent projects.

3. Virtual reality (VR) and augmented reality (AR): Students will be able to conduct experiments, visit art galleries or take trips into space in virtual classrooms or remote laboratories, even if they are not physically present. This technology will play a major role, especially in combining theoretical courses with practice.

4. Universities and diplomas

• **Digital and modular education:** Universities will offer modular, flexible and personalized education programs to their students. Instead of being tied to just one university, students will adopt a continuous learning model by accessing educational content from around the world.

• **Digital diplomas and accredition:** Instead of diplomas, achievements, skills and certificates will come to the fore in the learning process. Thanks to digital diplomas and block-chain-based accreditation systems, students' successes can be monitored transparently and securely.

• **Collaboration-oriented learning:** Universities will collaborate more closely with industry, providing students with opportunities to take part in real-world projects. Internships and work experience will become an integral part of university education.

5. Books and learning sources

- **Digital books and sources:** Traditional printed books will shift to digital formats. E-books, audiobooks, and interactive books will become more common. These books will provide students with more interactive content, providing a learning experience that is much more than just reading.

- **Personal learning platforms:** By having access to digital content and rich media resources (video, audio narrations, 3D models), students will be able to learn in a much more flexible way.

- **Costumazing trainable libraries:** Artificial intelligence will provide personalized book and resource recommendations for each student, so that each individual can easily find materials that suit their learning style.

6. Labs and experiments

- **Virtual labs:** Students will be able to conduct experiments in virtual laboratories without physically going to a laboratory. Thanks to VR and AR technologies, experiments in fields such as biology, chemistry and engineering can be conducted realistically in a virtual environment.

- **Robotics and automation:** In the laboratories, students will be accompanied by robotic lab assistants or AI-supported assistants. These assistants will guide students at every stage of the experiments and provide instant analysis and feedback on the results.

- **Online experience and hands-on training:** Real-world experiences will become more accessible through online platforms. Students will be able to gain hands-on experience with rapid prototyping, simulations, and machine learning through distance learning.

7. School closures and news education models

• **School will not close, they will evolve again:** Instead of schools closing completely, physical schools will integrate with technology, bringing more flexibility to the educational process. Physical schools will become places for more social interaction, creative learning and personal development.

• **Continuing education and lifelong learning:** Education will not be limited to school years. Throughout their lives, people will have access to different educational opportunities for specialization, retraining and personal development. Employees will take various certification programs and digital courses throughout their careers.

• **Mixed education experiments:** In addition to courses taken in traditional classrooms, students will have independent learning experiences on online and virtual platforms. Such flexible educational structures will transcend the physical boundaries of students' local schools.

In summary, education in the future will become more personalized, dynamic and interactive thanks to the integration of technologies such as digitalization, artificial intelligence, virtual and augmented reality. Schools and universities will not disappear completely; however, physical education places will transform into flexible and innovative centers where more digital technologies are integrated into teaching and learning processes. Face-to-face education and distance learning will coexist, and the culture of continuous education and lifelong learning will become more widespread.

KINDERGARTEN EDUCATION IN THE FUTURE

In the future, kindergarten education will take a very different shape than today due to technological developments, changing social needs and the evolution of pedagogical approaches. It is expected to be a system that focuses on developing children's individual talents, strengthening their emotional intelligence and helping them adapt to the world of the future. Here are some predictions about what kindergarten education may be like in the future:

Technology integrated education

• **Augmented reality (AR) and virtual reality (VR):** While learning, children can visit historical places, experience the workings of nature or explore the depths of space through virtual worlds. Such technologies will make the learning process more fun and interactive.

• **AI supported education:** Each child will be provided with individualized learning experiences that are tailored to their learning pace and interests. For example, an AI software can analyze a child's strengths and weaknesses to design activities specifically for them.

• **Robotic education:** Children can interact with robots to develop simple coding, problem-solving and creativity skills at an early age.

Focus on emotional intelligence and social skills

- **Empathy and emotional awareness training:** Kindergartens of the future will focus on developing children's emotional intelligence. Through drama, role-playing and group games, children will be taught the skills to empathize, express their feelings and communicate effectively with others.

- **Group projects and collaboration:** Teamwork and problem-solving skills will be a major focus for children to learn to cooperate.

Greater connection with nature

- **Outdoor education:** By spending more time in nature, children can learn about the cycles of nature, develop environmental awareness, and play creative games. Learning connected to nature will support both physical and mental health.

- **Agriculture and ecosystem trainings:** Small gardens can be set up in kindergartens and children can be taught practical information such as growing plants and getting to know the soil. This contributes to environmental awareness at an early age.

Inclusive and cultural education

- **Respect for diversity:** The kindergartens of the future will offer an inclusive learning environment that brings together children from different cultures, languages and lifestyles. Children will grow up learning respect and tolerance for differences at an early age.

- **Multi lingualism:** Learning foreign languages at an early age will become much more common. Children will be exposed to more than one language and will more easily embrace cultural diversity.

Education focused on creativity and problem solving

- **Activities that encourage creativity:** Children's creative thinking skills will be supported through activities such as art, music, dance and storytelling.

- **Game-based learning:** Play-based activities that will enable children to develop problem-solving and critical thinking skills will be at the forefront.

- **Health and well-being**

- **Mental and physical health education:** In the kindergartens of the future, practices such as mindfulness, yoga and meditation can be used to help children learn stress management and achieve emotional balance.

- **Nutrition and health awareness:** Interactive cooking workshops or how to make healthy snacks are taught to help children develop healthy eating habits at an early age.

Family and education partnership

- **Family involvement:** Kindergartens will involve parents more actively in education. Parents will be able to closely follow their children's learning processes and be a part of the education.

- **Online parent trainings:** Parents will be able to receive ongoing training on child development, educational methodologies and technology use.

Universal access and digital kindergartens

- **Digital kindergartens:** Especially in areas with transportation and geographic limitations, digital kindergartens can offer children the opportunity to receive quality education from home.

- **Equal educational opportunities:** In the future, more inclusive policies will be developed to ensure that every child has access to quality kindergarten education, regardless of their social and economic conditions.

The kindergartens of the future will offer an educational model where technology, nature and human-centered approaches come together, and where individual talents and emotional intelligence are at the forefront.

This system, which will support children's development not only academically but also in emotional, social and creative areas, will prepare them for the world of the future in a much more equipped way.

"Montessori education" will continue to have an important place in child development and education in the future. Some aspects of the Montessori method may change with technological developments, changing social structures and innovations in pedagogical approaches. However, its basic principles of respect for the child, individual learning pace, free discovery and learning through concrete experiences will be preserved. Here are the predictions for the future of Montessori education:

Balanced integration with technology

• **Interactive digital materials:** The traditional concrete materials of Montessori education (blocks, letter boards, math tools) can be supported by augmented reality (AR) and virtual reality (VR) in the future. Children can grasp abstract concepts more easily in a digital environment.

• **Screen time control:** Montessori's nature and concrete experience-focused structure can be adapted to ensure that children use technology in a balanced and limited way for education only.

Developing emotional and social skills

• **Emotional intelligence workshops:** Montessori schools may focus more on developing children's empathy, self-awareness, and communication skills. Mindfulness, meditation, and problem-solving games can be used in this process.

• **Community projects:** Montessori classrooms can include social responsibility projects that encourage children to interact with society at a young age.

Strengthening the connection with nature

• **Nature schools:** Montessori education emphasizes learning with nature. In the future, Montessori schools may transform into models such as "forest kindergartens" where children will spend more time outdoors.

• **Ecological awareness:** Practices that instill sustainable agriculture, recycling and environmental awareness in children may become more widespread.

Individualized education

- **AI supported tracking:** Each child's learning pace and interests can be monitored and analyzed with artificial intelligence systems. This data can guide teachers to create more individualized education plans.

- **Flexible learning models:** Montessori education encourages the child to learn at his or her own pace. This understanding can be expanded with greater flexibility in the future.

Cultural and multilingual education

- **Multilingual classes:** Montessori schools, which encourage language learning at an early age, may place greater emphasis on multilingual classes in the future.

- **Cultural awareness:** Programs where children from different cultures get to know and understand each other can become widespread.

Education that encourages creativity

- **Art and design focused workshops:** More emphasis will be placed on areas such as art, music and creative drama. Children's imagination and problem-solving skills will be supported.

- **Maker movement:** Montessori schools can be equipped with "maker" laboratories where children design simple engineering projects and solve problems.

Increased parental involvement

- **Parent education programs:** Montessori education continues not only at school but also at home. In the future, digital education platforms and home applications that comply with Montessori principles may be offered to parents.

- **Online tracking systems:** Parents can follow their children's development and learning processes through online systems.

Global reach and digital Montessori schools

- **Distance Montessori education:** To overcome geographical limitations, digital education platforms that comply with Montessori principles may emerge.

- **Global classes:** Montessori students from different parts of the world can come together through collaborative projects and digital platforms.

In the future, Montessori education will gain a more advanced structure with dynamics such as balanced integration with technology, individualized learning and return to nature. However, its basic values of free discovery, respect for the child and experience-based learning will continue to exist without change.

COMMUNICATIONS

With technological advances, digitalization, artificial intelligence and innovations such as augmented reality, communication methods will also undergo a radical transformation. How people communicate with each other will take on a much more layered, faster and interactive structure, especially when it comes to mobile phones and communication tools. Here are the developments expected with this change:

Revolution in audio-visual communication

• **AI supported communication:** Voice assistants will become more sophisticated AIs that respond to users' needs instantly. Phone calls will be managed by AI, and voice assistants will become smarter and provide instant answers to questions and queries.

• **Holographic conversations:** 3D hologram technology will enable people to talk to each other "in person" in a virtual environment, even if they are not physically together. Phones or smart devices can easily become devices that can make these holographic conversations.

• **Brain-computer interfaces (BCI):** It will be possible to communicate with brain waves. The brain-telephone connection will allow people to communicate with thoughts, words or even direct emotional states.

Augmented reality (AR) and virtual reality (VR) integration

• **AR glasses and lenses:** People will be able to interact with virtual worlds and other people with augmented reality glasses or contact lenses without the need for mobile phones. Text messages, social media posts, and emails will be visible on digital screens placed in the real world.

• **Virtual reality meetings:** In VR environments, people will be able to come together in virtual offices, virtual halls or virtual homes, and will be able to carry out all kinds of social interactions in a 3D virtual world. Such meetings will redefine personal connections by imitating real-world interactions.

Biometric and emotional communication

• **Emotional communication techniques:** People will also be able to share their emotional states through technological tools. Biometric sensors or artificial intelligence will understand a person's mood and provide appropriate responses. For example, a phone that understands a friend's mood will send appropriate messages to them or will perceive their emotional state from their tone of voice and provide more empathetic responses.

• **Emotional facial expressions:** People will be able to see the emotional state of the other person not only from texts but also from their digital avatars or holograms. In addition, content sent on social media will become more meaningful with the analysis of emotional state.

Super fast communication with 5G, 6G and beyond

• **Delay-free communication:** With 5G and 6G, communication speeds will become much faster, enabling instant responses and seamless video, voice and holographic conversations.

• **Fast data sharing:** File sharing, large data transfer and media file communication will be done quickly, online content will become more dynamic. People will be able to share 3D and VR content instantly

Smart devices and IoT integration

• **Communication between devices:** A continuous communication network will be established between smart devices. People will communicate by switching between their phones, smart devices at home, wearable technologies and vehicles. For example, you will be able to reply to a message from your phone with your smart watch or smart screen at home.

• **Digital assistants and IoT:** People will make their lives easier by communicating with digital assistants from their homes, vehicles and workplaces. Smart homes will analyze users' daily needs and offer them appropriate communication solutions.

Personal and global communication

• **Digital nomadism and time zone issues:** People will come together in different villages or cities around the world and adopt a global lifestyle by communicating. Uninterrupted and global communication will be possible at any time, regardless of time zones.

• **Digital identities and avatars:** People will interact with their digital identities and avatars, which will behave according to their personal characteristics, skills, and preferences, and even represent their true selves on social media and in virtual worlds.

Translation systems that remove language barriers

• **Real-time translation:** Language barriers will be overcome with instant translation technologies. People will be able to communicate in their natural language and instantly respond in another language.

• **Comprehensive and personalized translations:** Smart devices will make global communication more efficient by making translations more accurate and personalized based on a person's language proficiency level.

In summary; the way people communicate in the future will be faster, more personalized and more interactive. Holograms, smart devices, biometric data and virtual worlds will allow people to interact with each other more deeply and realistically. With these opportunities offered by technology, communication will mean much more than just voice or text.

EVOLUTION OF SOCIAL MEDIA PLATFORMS

Yes, it is likely that social media networks that are very different from existing platforms will emerge in the future. Developments in technology, changing user expectations, and diversifying digital experiences will shape these new platforms.

Possible new types of social media platforms

1. Metaverse-based social networks

• Users will interact with each other in virtual reality (VR) or augmented reality (AR) environments.

• Friendships, meetings and events will be organized through virtual avatars beyond physical presence.

• Virtual concerts, parties and shopping experiences will be a part of social media.

2. Emotion and biometric based platforms

• Platforms will analyze users' emotional state (happy, stressed, sad) and provide appropriate content.

• Customized interactions will be provided with biometric data (e.g. content recommendation based on heart rate).

3. AI personal assistant networks

• AI assistants specific to each user will provide support from content creation to link suggestions.

• Assistants can automate interactions by communicating on behalf of the user.

4. Anonymous and decentralized networks (Web3 social media)

• Blockchain-based platforms will be independent of central controls.

• User data will be completely personal property and content can be shared in NFT format.

• Communities focused on freedom and privacy can develop.

5. Time-limited content sharing platforms

• Shared content will be automatically deleted after a certain period of time.

• Users may gravitate towards a social media culture focused on "instant experiences."

6. Sound and hologram based platforms

• Platforms dominated by voice chat will develop further (like Clubhouse).

• Real-time interactions with holographic images may become popular.

7. Vertical interest platforms (niche networks)

• Networks that focus solely on specific interests (e.g. arts, science, health, book communities).

• Users will be able to connect only on topics that interest them, breaking away from the chaos of general social media.

8. Networks focused on ethics and mental health

• Platforms that prioritize protecting users' mental health.

• Communities that block negative content and focus on positive interactions.

9. Digital human and AI interactive platforms

• Platforms where users can become friends not only with humans but also with artificial intelligence.

• Communication can be established with digital avatars, virtual partners or chatbots

10. Mixed reality interaction platforms (xr media)

• Platforms where the physical world and the digital world are completely intertwined.

• Digital content can be shared instantly with augmented reality glasses while walking on the street.

In the future, social media platforms will be more personal, experience-oriented, free, and technological. We will see networks where users not only consume content but also actively participate in experiences and even play a role in the design of the platforms. Privacy, personalization, and virtual reality will be at the center of this evolution.

THE FUTURE OF SOCIAL MEDIA

Technological developments, changing user expectations and digital trends may cause some social media platforms to lose their popularity over time and face the risk of extinction. Below are the factors that may affect this situation and the platforms that may be in danger.

1. Centralized and data driven platforms (Examples: Facebook, X/Twitter)

Why is it in danger?

• User privacy violations and data security concerns.

• Younger users moving away from centralized platforms.

Potential future

• Centralized structures may be replaced by blockchain-supported social networks where user data is under control.

2. Short-term video platforms (Example: TikTok)

Why is it in danger?

• Users' attention spans are getting shorter due to the fast consumption culture.

• Regulations and restrictions on data collection policies.

• New types of interaction (VR/AR) are leaving video behind.

Potential future

• Virtual reality and metaverse-focused content platforms can replace short videos.

3. Photo and visual-oriented networks (Example: Instagram)

Why is it in danger?

• Saturation and content repetition.

• Popularization of more immersive visual experiences in AR/VR environments.

• Users gravitate towards more unique and temporary content.

Potential future

• Platforms like Instagram may not be able to compete without holographic content or augmented reality integrations.

4. Text-oriented platforms (Examples: Reddit, X/Twitter)

Why is it in danger?

• The rise of audio content and holographic communication technologies.

• The decline of users' interest in long texts.

Potential future

• Voice and augmented reality chat networks can replace text-oriented social networks.

5. LinkedIn style professional platforms

Why is it in danger?

• Direct integration of artificial intelligence-supported job search algorithms and individual digital identities in the professional environment.

Potential future

• Blockchain-based digital career identities may come to the fore instead of centralized platforms.

6. Streaming and live broadcast networks (Examples: Twitch, YouTube)

Why is it in danger?

• Users are turning to fully interactive content.

• Virtual reality broadcasts are changing the viewer experience.

Potential future

• "Immersive" experiences where users actively participate in broadcasts instead of passive spectators.

Conclusion: Platforms that cannot adapt to the future will disappear

• **Adapters:** Platforms that keep up with trends such as the metaverse, holographic communication, AR/VR integration, and data privacy will continue to exist.

• **Those that might disappear:** Social media networks that remain dependent on old technologies and centralized structures and fail to renew user experience will be doomed to disappear in the future.

In the future, the platforms that drive the digital world will be based on more dynamic, participatory and user-centered systems.

EVOLUTIONARY CHANGE OF MEDIA JOURNALISM

News has been an inseparable part of social life throughout human history. However, journalism has become a constantly changing and transforming field with the development of technology. In the future, the way news is produced, disseminated and covered in the media will be very different from today and will enter a completely new era with technologies such as artificial intelligence, blockchain, augmented reality (AR) and virtual reality (VR). In this book, we will discuss how news will be created in the future, how it will be disseminated and what transformations journalism will undergo.

Artificial intelligence and automation in news production

In the future, much of the news writing, editing and content creation process will be automated. Even today, some agencies are doing AI-powered news writing, but this process will become much more advanced in the future.

1. AI-supported journalism

- **Data driven news:** Artificial intelligence will be able to quickly interpret economic, political and social developments by analyzing big data. For example, events such as sudden changes in the stock market or election results will be processed instantly and turned into news.

- **Real-time content production:** Algorithms will be able to collect instant information from social media and other news sources, analyze events instantly and write news.

2. Language model supported editing: The consistency of the news in terms of language and expression will be ensured by artificial intelligence systems, thus reducing the burden on human editors.

3. Robot journalists and automated reporters

In the future, unmanned news reporters will go into the field.

• **Drone and robot reporters:** They will instantly capture images from the scene, analyze them and transfer data to news centers.

• **Virtual reporters:** Virtual news anchors created by artificial intelligence will be able to present the news in real time.

However, it should not be overlooked that artificial intelligence may create ethical problems and increase the risk of the spread of manipulated news.

How news spreads: New media and social platforms

Although traditional media (newspapers, television, radio) will still exist, news in the future will become individualized and focused on instant consumption.

1. Personalized news feed

• **AI supported news filtering:** News will be automatically sorted and presented according to users' interests.

• **Smart assistants:** Google Assistant, Siri or future artificial intelligence-supported systems will be able to create newsletters tailored to individuals.

2. Social media and instant broadcasting

• **Individual journalism:** As is the case today, individuals will continue to practice citizen journalism by sharing news directly on social media platforms.

3. The power of live broadcasts: Platforms such as TikTok, Instagram Live and YouTube will further develop and enable instant news transmission.

4. Holographic and 3D news

In the future, news could become a three-dimensional experience with augmented reality (AR) and virtual reality (VR).

- **Virtual press conferences and interviews:** Journalists will be able to conduct virtual interviews anywhere in the world using hologram technology.

- **Experiencing the news live:** It will be possible to follow the news by directly experiencing the developments in a war zone or a natural disaster with VR glasses.

The future of magazine and current news

The current news and magazine world will also undergo a major change.

1. Virtual celebrities and digital magazines

- **Digital influencers:** Today, there are AI-powered virtual phenomena like Lil Miquela. In the future, completely virtual celebrities and digital characters will shape popular culture.

- **Personal brand management:** Celebrities will manage their own images with algorithms that can automatically control how news spreads about them.

2. Risk of deepfake in gossip and magazine content

- **Deepfake videos:** Fake videos that show celebrities saying things they didn't say can cause major controversy in the magazine world.

- **Fake scandals:** It will become much easier to create fake scandals through the manipulation of algorithms.

At this point, media ethics will become even more important and fact-checking systems will have to improve.

The future of media: how will news agencies and broadcasting evolve?

Traditional media will not disappear, but will have to keep pace with the digital transformation.

1. Blockchain and trusted news broadcasting

• **Encryption of news:** Blockchain technology will be used to prevent fake news by verifying the authenticity of news.

• **Independent and decentralized journalism:** Instead of being controlled by large media outlets, news can be shared by individual journalists on independent blockchain-powered platforms.

2. AI-powered editors and algorithms

• **Unbiased journalism:** AI-powered editors who measure the accuracy and impartiality of news can be used to ensure the impartiality of media organizations.

• **Automatic content distribution:** Media organizations will be able to disseminate news directly through algorithms without the need for human editors.

Ethical and legal dimensions of journalism in the future

In such a rapidly developing news world, ethical and legal regulations will become inevitable.

• **Fake news and manipulation:** New laws and filtering systems will be developed to prevent AI-supported news from being used for manipulative purposes.

• **Privacy and data usage:** Presenting news according to users' interests may lead to misuse of personal data. Therefore, data privacy policies will gain importance.

• **Censorship and free press:** Whether AI-powered media algorithms should block certain content will be a major ethical debate.

How will the future of journalism be shaped?

• News will become faster, more instantaneous and more personalized.

• Artificial intelligence will be present at every stage from news writing to editing.

• Traditional media will not disappear, but it will be completely digitalized.

• Augmented reality and holographic technologies will transform news into a more interactive experience.

• Ethics and reliability will continue to be the biggest topics of discussion.

In the future, journalism may become both freer and more controlled with technological advances. However, the biggest question is: How will real and fake news be distinguished, and how impartial can journalism remain? Technological advances, artificial intelligence, digital media platforms and social media are rapidly changing the way we receive news.

In the future, people will receive news in a more personalized, interactive and real-time way. Here are the important transformations that will occur with this change:

Personalized news experience

• **AI and machine learning:** Users will encounter newsletters and content customized to their personal preferences. Artificial Intelligence will present news based on individuals' interests and previous reading habits.

• **Instant and dynamic update:** News will be distributed in real time according to people's interests and geographic location. For example, events in a city will be instantly presented to the user via a map and video integration.

1. Interactive and video-base content

• **Video and live broadcasts:** News will be presented mostly in video format. Users will be able to watch not only written content but also interactive live broadcasts, video interviews and live news feeds.

• **Augmented reality (AR) and virtual reality (VR):** The news will offer viewers a virtual news viewing experience, with mixed reality technology bringing the events to life in person.

- **Live interactive platforms:** Users will be able to instantly interact with news presenters and ask questions or participate in discussions.

2. The role of social media and crowd-sourced journalism

- **The power of social media:** Instead of traditional media outlets, social media platforms will become more prevalent in disseminating news. Users will create their own "news networks," going directly to social media influencers and independent journalists for accurate and fast information.

- **Crowd-sourced journalism:** People will share instant news with mobile devices and smart glasses. Everyone will be able to become a news source through digital platforms.

- **Verification and reality checking:** Artificial intelligence will try to prevent the spread of false information by instantly checking the accuracy of news spread on social media.

3. Subscribtion and on-demand news services

- **Special subscribtion models:** Users will subscribe to platforms that offer in-depth news content in a particular niche (sports, science, politics, etc.).

- **On-demand news feed:** News, just like music or video streaming services, can be watched at any time, depending on timing and genre selection.

- **Personal news anchors:** Each user will be assigned a personal "news assistant" who will collect and present news according to the user's wishes.

4. Transformation in magazine and current news

- **More inclusive and diverse magazine content:** Magazine news will no longer be limited to celebrities; rich content will also be offered on culture, lifestyle, environment and social issues.

- **Interactive magazine programs:** Viewers will take an active role in magazine programs. For example, viewers will be able to directly participate in the broadcast and influence the content.

- **Digital image and interactive lives:** Celebrities will be able to interact with their digital avatars in virtual environments and news will be presented through these avatars.

5. Digital and hyper-micro content consumption

- **Fast consumption:** News will be presented in a short and concise format. TikTok and similar platforms will convey news in a few seconds of video.

- **Micro-news:** Short, fast and easily consumable content will enable users to obtain information very quickly.

6. Holographic and virtual news presentations

- **Holographic presentations:** With 3D holograms, news anchors will present the news to viewers without being physically visible.

- **Events experienced with virtual reality (VR):** Users who want to be at the center of events will be able to watch the news "live" with virtual reality glasses.

In the future, media and news will become more personalized, faster, interactive and visually focused. Users will not only watch the news, but will also participate in content production, verification and interaction processes. This transformation signals a period in which digital platforms will gain greater power alongside traditional media outlets and a "participatory" media experience will come to the fore.

CINEMA - TV SERIES PRODUCTION

Yes, film and series production with artificial intelligence will become possible in the future, and this process has already started. The use of artificial intelligence (ARTIFICIAL INTELLIGENCE) in the cinema and series industry will create major transformations in both creative and technical aspects.

Film and series production with artificial intelligence

1. Script writing

Artificial intelligence can create new stories by analyzing thousands of previously written scenarios. With the collaboration of creative writers, AI can provide faster and more effective starts to scenarios.

2. Virtual actors (digital avatars)

• Fully digital characters can be created without the need for real actors.

• Digital twins of existing actors can be created and they can be aged, rejuvenated, or even "played" after their death.

3. Visual effects (VFX) and animation

AI enables complex scenes to be rendered quickly and realistically. For example, massive battle scenes can be simulated without human intervention.

4. Voiceover

Artificial intelligence can create fully digital voice-overs by imitating any person's voice.

5. Directing and editing

AI can make suggestions or auto-edit scene composition, camera angles, and scene transitions.

Identifying artificial actors

1. Digital versions of real actors: Artificial intelligence can create digital copies of actors by analyzing their facial expressions, voices, and movements. This way, the same actor can play in multiple projects "at the same time."

2. Completely artificial characters: Virtual actors created from scratch, not based on any human being, could become popular. These characters can be optimized for the audience (e.g., faces and behaviors adapted to different cultures).

3. Optimization with audience data: By analyzing audience data, AI can predict which types of actors and stories will be more popular and shape projects accordingly.

Possible advantages

• **Cost savings:** Actors, sets and shooting costs can be greatly reduced.

• **Fast production:** Production can be completed in a much shorter time.

• **Infinite creative possibilities:** Desired scenes and characters can be created without any physical or logistical limitations.

Possible problems and ethical debates

1. Job loss: Reduced job opportunities for actors, screenwriters, and other creative professionals.

2. Copyright: Copyright and permission issues related to the use of digital copies of actors.

3. Identity issues: Confusion of artificial characters with real people and "deepfake" issues.

4. Robotization of creativity: Lack of human emotion and creative intuition.

As a result, AI-supported film and series production may become standard in the near future. However, this transformation will require redefining creative processes and establishing ethical frameworks.

THE FUTURE OF MUSIC NEW ERA MUSIC GENRES

The music industry is rapidly transforming under the influence of technology, artificial intelligence and digital platforms. In the future, the music listening experience will become more personal, dynamic and participatory. Traditional music genres will be intertwined with innovative music approaches and the listener will be offered completely unique experiences.

Personalization and AI in music

• **Compositions with artificial intelligence:** Artificial intelligence algorithms can create instant musical compositions based on an individual's mood and listening habits.

• **Mood-based music:** Music suggestions will be made automatically by analyzing the listener's emotional state (happy, sad, energetic).

• **Biometric responses:** Smart devices can provide personalized music production by analyzing heartbeat and brain waves.

• **Holographic concerts:** Hologram concerts of artists who lived in the past can be organized with virtual reality technology.

The future of traditional and modern music genres

Pop music

• AI-supported "hit" songs will be produced rapidly.

• With platforms like TikTok, popularity will be more short-term and dynamic.

Arabesque

• Arabesque music will be reinterpreted with more electronic substructures and will evolve into genres such as "Arabesque-Trap".

• It will be possible for personal pain and sadness themes to be supported by individualized compositions by artificial intelligence.

Folk and art music

• Turkish Folk Music and Turkish Classical Music will be transferred to the new generation via digital platforms.

• Traditional instruments will be combined with modern electronic infrastructures (for example, EDM with bağlama).

Turkish music

• Classical Turkish music will be revived with virtual orchestra projects.

• Cultural heritage will be preserved by digitizing through academic studies.

Rap and hip-hop

• Rap will continue to be the most powerful music genre that continues social criticism.

• With artificial intelligence, individual stories can be transformed into rap pieces specially composed for listeners.

Rock

• Rock music will evolve with digital and electronic infrastructures.

• VR concerts and virtual guitar performances may become popular.

Elektronic music

• Electronic music automatically produced by artificial intelligence will become widespread.

• DJs may be replaced by artificial intelligence music systems.

Future technologies in music

• **NFT music:** Artists will sell their music as digital assets and offer collectors unique experiences.

• **Metaverse concerts:** Interactive concert experiences will become common in virtual reality universes.

• **Copyrights with Blockchain:** Artists will be able to protect their works more securely and transparently.

Music culture of the future

Traditional music genres will not disappear; however, they will be reinterpreted by merging with digital and technological infrastructures. As people's connection with music becomes more personal, strict boundaries between genres will disappear. Both nostalgia and innovation will live together.

X

FUTURE ENVIRONMENTAL
POLICIES AND RELATIONSHIP
WITH NATURE

SOLAR RADIATION CONTROL / SPACE MIRRORS AND SHADOWS

It is not possible to completely control the sun's rays reaching the Earth with current technology, but there are some studies and theories in this direction. In particular, solar engineering (solar geoengineering) and space-based solar energy projects bring up the possibility of partially directing or controlling sunlight in the future.

Potential control mechanisms

1. Stratospheric aerosol injection (SAI)

• This technique aims to reflect some of the sunlight back into space by releasing aerosols such as sulfate particles into the atmosphere.

• Although it has come to the fore with the idea of controlling global warming, the possibility of certain powers abusing this technology is a controversial issue.

2. Space mirrors and shades

• Theoretically, it might be possible to direct or block sunlight to certain areas using large space mirrors or shades.

• Such a system raises the question of whether it could be used as a geopolitical weapon to warm or cool certain countries.

The future of climate control and energy management may be in the hands of certain powers.

3. Space-based solar power (SBSP)

• The idea of collecting energy with giant solar panels and transmitting it to Earth via microwaves or lasers has been researched for a long time.

• If this system is in the hands of major powers, a serious monopoly on energy supply could be created.

Could solar energy be a power tool in the future?

• Depending on technological developments, some countries or companies may have the capacity to control sunlight.

• Energy monopolization may create a new global balance as solar energy comes under the control of certain powers.

• Its use as a geopolitical weapon may create economic and political pressure by driving one region into drought or providing too much light to another.

For now, controlling the sun's rays completely seems like science fiction, but it is not out of the question that some countries or companies may gain the upper hand in this area with the development of such technologies in the future. How do you think this situation may affect the world balance?

**Space mirrors and shades:
The future of climate control and energy management**

Space mirrors and sunshades are innovative engineering projects designed to redirect or block sunlight. These concepts have a wide range of uses, from combating climate change to space-based energy production and even geopolitical weapons systems.

The basic logic of space mirrors and shades

Space mirrors and sunshades can be thought of as giant structures placed between the Earth and the Sun. These structures aim to control the amount of heat the planet receives by reflecting or redirecting some of the sun's rays back into space. This idea can serve two basic purposes:

1. Combating climate change

• A certain percentage of the sun's rays coming to Earth can be blocked to reduce global warming.

• Overheated regions can be cooled.

2. Energy management and use

• Energy production can be increased by concentrating solar energy in certain areas.

• It can be integrated with space-based solar panels.

3. Working principle of space mirrors

• Space mirrors are systems that focus or disperse sunlight to a specific point using large-scale reflectors. Theoretically, it works as follows:

• **Low-orbit mirrors:** Large mirrors orbiting the Earth can support agriculture and energy production by reflecting sunlight to desired areas.

• **Lagrange point (L1) sunshades:** A giant mirror or shade can be placed at L1, one of the gravitational balance points between the Earth and the Sun, to adjust the amount of light reaching the Earth.

The design of such systems requires ultra-light materials, automatic placement technologies, and nanotechnology coatings that will reflect sunlight in a controlled manner.

Space shades: Is it possible to block out the sun's rays?

Space shades could act like giant sun umbrellas, changing the temperature balance by shading certain areas.

• **Climate control:** To reduce global warming, a certain percentage of the sun's rays falling on the Earth can be targeted.

• **Regional cooling:** It can make agriculture more sustainable by reducing extreme temperatures in arid regions.

• **Planetary impact:** Blocking 1-2 percent of the sunlight reaching the Earth may be enough to reduce global temperatures by 1-2°C.

Such projects would involve deploying large canopies in space made of super-thin, ultra-light materials. Theoretically, these systems could be stabilized with micro-solar sails and magnetic fields.

Technological and engineering challenges

The construction of space mirrors and shades is an extremely complex process and presents many technological challenges:

a. Material and construction difficulties

• The materials to be used must be light, durable and highly reflective.

• How will large structures be built in space? Will foldable and automatic opening systems be used?

b. Positioning and stabilization

• These structures must be placed at precise points between the Earth and the Sun and remain balanced there.

• Factors such as orbital changes and solar winds require constant adjustments for stabilization.

c. Financial costs

• Placing such large structures in space could cost trillions of dollars.

• Would governments and private companies invest in such a project?

4. Potential areas of use

a. Preventing global warming

• The idea of cooling the Earth by reflecting some sunlight is seen as a solution to combating climate change.

• Sulfate particles released into the atmosphere as a result of the 1991 eruption of Mount Pinatubo lowered global temperatures by 0.5°C. Space canopies could have a similar effect.

b. Space-based solar energy production

• Solar energy collected in space can be sent to Earth in the form of microwaves or lasers.

• Reflective mirrors can be used to create areas that receive sunlight even at night.

c. Agriculture and economic use

• Agricultural productivity can be increased with specially directed sunlight for arid regions.

• It may be possible to direct sunlight directly to power plants to increase energy supply.

5. Dangers and ethical issues

a. Geopolitical risks

• If a country or company took control of these systems, could they use the climate as a weapon?

• Could directing or blocking sunlight to certain areas become a weapon of war?

b. Side effects

• Interfering with the Earth's natural climate balance could lead to unexpected weather events.

• There could be possible negative effects on ecosystems and agriculture.

c. Space pollution and collision risks

• Such structures can increase satellite traffic and space debris, leading to dangerous collisions.

Science fiction or reality

Although space mirrors and shades are currently in the conceptual stage, they may become applicable in the future with the development of space technologies. NASA, the European Space Agency (ESA) and some private companies are theoretically investigating such projects.

However, such technologies also bring with them a great deal of ethical and geopolitical debate. If space mirrors and shades are ever implemented, one of the biggest questions will be who will control them.

Although the issue of controlling solar radiation in the future may seem like science fiction, it may become a serious area of debate when looking at the course of climate change and energy policies. In this context, controlling solar radiation may play a critical role in both energy production and the planet's climate balance.

271

TECHNOLOGIES TO CONTROL SOLAR RADIATION

1. Solar energy satellite systems (Space-Based Solar Power - SBSP)

Giant solar panels placed in space can be used to collect sunlight directly and send energy to Earth. When this technology is developed, some countries or private companies can establish a large monopoly on energy.

2. Solar geoengineering

There is an idea to reflect some of the sun's rays by releasing substances such as sulphate aerosols into the atmosphere. This has been proposed to slow global warming, but carries ecological and political risks.

3. Atmospheric mirrors and reflector systems

It is thought that sun rays can be directed to or blocked from certain areas using giant mirrors or space-based reflectors.

4. Control of photovoltaic infrastructure

The management of the world's large solar energy fields and grid systems will also be an important part of energy policies.

Who will be in control in the future?

• **States and süper powers:** Major powers such as the United States, China, and the European Union are likely to be leaders in solar energy infrastructure. Space-based solar energy systems could become part of the strategic plans of major powers.

- **Global organizations:** The United Nations or similar global organizations could take on a supervisory role in geoengineering projects such as solar radiation control.

- **Private companies:** Space technology companies such as SpaceX, Blue Origin and others could make major investments in solar energy satellite systems and become strong players in the energy sector.

- **Geoengineering consortia:** International consortia formed by scientists and technology giants can manage sunlight intervention projects for climate balance.

Ethical and political issues

- **Monopoly and energy hegemony:** Control of energy provides economic and political advantage. Manipulation of sunlight can increase this power even further.

- **Climate justice:** Reducing sunlight can cause some areas to cool and others to become dry.

- **Security risk:** The use of geoengineering projects by malicious groups can lead to ecological disasters.

In summary; the control of solar radiation can become a major strategic issue in both energy and climate policies. Therefore, who will have this power in the future will remain a critical question in terms of global security and justice.

ARTIFICIAL CLOUDS ARTIFICIAL RAIN AND SNOW

Climate change technologies: Is it possible to control rain, snow and drought?

Throughout history, humanity has desired to control the climate. While in the past, attempts were made to influence nature through rain prayers, dances, and various rituals, today science and technology are taking concrete steps to achieve this goal. Is it possible to make it rain and snow in certain regions, or even make long-term climate changes, through methods such as cloud seeding, stratospheric interventions, and atmospheric engineering? Today, the use of such techniques for military and political purposes has also become a frequently debated topic.

Cloud seeding: Is it possible to produce rain and snow?

1. Basic principle of cloud seeding

Cloud seeding is the process of adding chemicals to clouds to increase rain or snowfall. The most common substance used is silver iodide, because it encourages ice crystals to form on water droplets.

How does it work?

Silver iodide, salt or dry ice (carbon dioxide) are sprayed into the clouds by airplanes or rockets. These substances cause the water vapor in the cloud to condense, forming rain or snowflakes. By increasing precipitation, arid regions can be provided with water.

Success rate

Scientific studies have shown that cloud seeding can increase precipitation by 10-20 percent. However, it is not currently possible to create clouds completely artificially; only existing clouds can be manipulated.

Which countries use it?

China: Used cloud seeding to prevent rain during the 2008 Beijing Olympics.

United Arab Emirates: Carries out major projects, allocating millions of dollars to rain in deserts.

USA: Conducts cloud seeding to combat drought in agricultural areas.

Russia: Uses this technology to prevent rain during major celebrations.

2. Climate control: is it possible to create drought?

Rainfall can be increased with cloud seeding, but is it possible to drive a region into drought? Theoretically, it may be possible to create drought in certain regions with certain methods.

• **Attacking use of cloud seeding:** If rain clouds in a certain area are artificially directed to another area, an artificial drought can be created there.

• **Atmospheric heating systems:** By heating certain layers of the atmosphere with high-frequency radio waves, rain formation can be prevented.

• **Artificial sun screens:** Regional temperatures can be changed by increasing or decreasing sunlight.

Throughout history, it has been claimed that some countries have explored atmospheric engineering techniques to create drought in enemy territory.

Can climate change be done?

Beyond short-term weather changes, is it possible to create long-term climate change?

a. Stratospheric Aerosol Injection (SAI)

This method aims to block sunlight by spraying sulphate particles into the upper atmosphere. Large volcanic eruptions have a similar effect. For example, the 1991 eruption of Mount Pinatubo lowered global temperatures by 0.5°C. If this technique were applied on a large scale, it might be possible to slow global warming.

b. Ocean fertilization

The aim is to increase plankton production by adding chemicals such as iron sulphate to the oceans. Plankton can reduce global warming by absorbing carbon dioxide from the atmosphere.

c. Blocking the sun rays (space shades)

By using giant mirrors or artificial curtains, the amount of sunlight reaching the Earth could be reduced, which could lead to the planet cooling down.

Could climate engineering be used as a weapon?

Climate manipulation can be done for peaceful purposes, but it also has military and geopolitical implications. Since the 1960s, some countries have been exploring the potential of using weather and climate as tools of war.

a. HAARP and atmospheric interventions

• The US HAARP Program is a project that researches changing weather conditions by heating the atmosphere with high-frequency radio waves.

• According to conspiracy theories, HAARP can create earthquakes, hurricanes or artificial droughts. However, there is no scientific evidence for these claims.

b. Desertification and wars over water resources

• If a country attracts rain clouds or creates droughts from a rival country, this can lead to economic collapse.

• Climate change can increase migration and cause political crises.

c. Artificial hurricanes and storms

• Some theories suggest that artificial hurricanes can be created by controlling the temperature and air pressure in large seas.

• However, to do this properly would require enormous energy and such technology does not yet exist.

Ethical and legal issues

Although climate engineering projects have great potential, they also come with many ethical and legal problems.

• Which country has the right to intervene in the climate to what extent?

• Can unconscious interventions lead to unexpected disasters?

• If these technologies are monopolized as a tool of power, could a new type of colonialism emerge in the world?

The United Nations signed an agreement under the name of "Convention Prohibiting the Use of Environmental Modification Techniques for Military or Any Hostile Purposes" (ENMOD) in 1978. However, it is unclear how much such technologies can be controlled.

Conclusion: Will weather and climate be controllable in the future?

• Rainmaking and snowmaking technologies are partially possible today.

• Creating droughts and long-term climate change are more complex, but some techniques are theoretically available.

How these technologies will be used in the future is a matter of great ethical and political debate.

If humanity were to gain full power to change the climate, would these technologies save the world, or would they cause a great disaster?

Artificial clouds and artificial rain/snow technologies

Technologies for creating artificial clouds and manipulating the atmosphere to make it rain or snow have become an important area of research in recent years in the fight against climate change and water scarcity. This method, known as "cloud seeding," is based on encouraging precipitation by intervening in the composition of natural clouds.

How to create artificial rain and snow?

1. Cloud Seeding

This method is carried out by sprinkling chemicals (usually silver iodide, salt crystals or dry ice) into the clouds via airplanes or unmanned aerial vehicles. These substances facilitate the condensation of water vapor, thus accelerating precipitation.

• **Artificial rain:** Thanks to seeding, water droplets come together and cause rain to fall.

• **Artificial snow:** In cold weather, snowfall can be triggered by creating ice crystals through the seeding process.

2. Atmospheric moisture condensers

Some high-tech devices can produce water directly by condensing moisture in the atmosphere.

3. Artificial cloud generation

Studies are ongoing on systems that provide moisture condensation by heat and steam production. In these systems, water vapor can be raised and cloud-like structures can be created with certain chemicals.

Areas of use and potential benefits

1. Fighting drought: Artificial rain can be provided for agricultural activities in arid regions.

2. Water management: Rainfall enhancing techniques can be used to increase the filling rates of dams.

3. Weather control: On special events or critical days, rain may be avoided or snowfall may be encouraged.

4. Fire extinguishing: During forest fires, artificial rain can be used.

Countries where artificial rain and snow are used

China: During the 2008 Beijing Olympics, rain clouds were seeded to keep the air clean.

United Arab Emirates: Artificial rain projects are actively implemented in the desert environment.

ABD: Cloud seeding has been used in agricultural areas for many years.

Rusya: Cloud seeding technology is being used in Moscow to prevent rain on special days.

Ethical and ecological issues

• **Ecological balance:** Interference with the atmosphere can disrupt natural precipitation cycles.

• **Water rights:** It may create conflicts over cross-border water resources.

• **Climate manipulation:** In case of misuse, it can lead to major environmental disasters.

• **Monopolization of air control:** Monopolization of these technologies by certain countries or companies may pose a risk.

In summary; artificial clouds and artificial precipitation technologies can be an important solution in combating drought and climate change. However, these technologies should be used within ethical frameworks without disrupting the ecological balance.

XI

ARTIFICIAL INTELLIGENCE AND GLOBAL POLICIES

CAN ARTIFICIAL INTELLIGENCE BE BANNED?

Whether or not AI will be banned in some countries depends on many factors. This possibility can be discussed, especially in developing or third-tier countries with authoritarian governments. Here are the prominent evaluations on this subject:

1. Possibility of getting banned

• **Information control:** Authoritarian regimes may be uneasy about AI's ability to increase access to information and strengthen freedom of expression, so AI technologies may be banned or tightly regulated.

• **Competition concerns:** Countries that cannot keep up with technological developments may block foreign AI products and services on security grounds.

• **Security concerns:** Cybersecurity threats and data security concerns may lead to restrictions on AI.

2. Regulation instead of prohibition

Rather than a complete ban, the following regulations are more likely in these countries:

• Local data protections: Limits could be placed on the activities of foreign AI companies.

• Content moderation: AI applications could be subject to censorship.

• Licensing: Only AI products authorized by the state could be encouraged to be used.

3. Economic and social impacts

• **Technological backwardness:** Banning AI could further reduce the economic competitiveness of these countries.

• **Education and employment issues:** Being deprived of the training and automation opportunities offered by artificial intelligence can hinder development.

A complete ban would not be a practical solution. However, authoritarian regimes can keep AI under tight control to keep information and technology under control. In order for developing countries to integrate this technology, conscious regulations and educational investments are critical. Artificial intelligence has an important place in the strategic plans of countries today. Each country develops AI policies in line with its own priorities and resources. These policies are shaped according to the economic, social and technological goals of the countries.

COUNTRIES' APPROACHES TO AI

Below, some countries' perspectives and strategies on artificial intelligence are examined in detail:

The United States

The United States, which is a leader in the field of artificial intelligence, is developing comprehensive strategies to maintain this leadership. The document titled "Maintaining American Leadership in Artificial Intelligence" published in 2019 aimed to prioritize the research and development of artificial intelligence, establish ethical standards, and strengthen international collaborations. In addition, a comprehensive decree was issued in 2023 to prevent risks arising from artificial intelligence.

China

China has placed artificial intelligence at the center of its national strategy and published the "Next Generation Artificial Intelligence Development Plan" in 2017. With this plan, it aims to become a world leader in artificial intelligence by 2030. China sees artificial intelligence not only as an economic tool, but also as a national security and social management tool. In this context, it is expanding artificial intelligence applications using big data and surveillance technologies.

The European Union

The EU adopts a human-centered and ethical approach to AI. The "Economic Impacts of Artificial Intelligence" report, published in 2019, emphasized the contribution of AI to economic growth, while also warning about the negative effects it could have on the labor market. The EU is developing guidelines and regulations for the ethical use of AI, while also encouraging cooperation between member states.

United Arab Emirates (UAE)

The UAE is among the countries that attach great importance to artificial intelligence. It has demonstrated its determination in this field by establishing the world's first Ministry of Artificial Intelligence. It also aims to become a regional AI hub by investing in AI education and research.

Singapur

Singapore, which sees artificial intelligence as the key to economic development, is developing comprehensive strategies in this area. It aims to train a skilled workforce, especially by placing emphasis on artificial intelligence education. It also adopts policies for the ethical and safe use of artificial intelligence.

France

France is developing strategies to be among the leading countries in the field of artificial intelligence. It aims to increase its capacity in this field, especially by investing in artificial intelligence research and development activities. It is also creating policies for the ethical and safe use of artificial intelligence.

Turkey

Turkey has revealed its vision in the field of artificial intelligence by publishing the "National Artificial Intelligence Strategy" in 2021. With this strategy, targets such as training artificial intelligence experts, supporting research and innovation, and expanding access to quality data and technical infrastructure have been determined. In addition, it is planned to make regulations for the ethical and safe use of artificial intelligence.

Germany

Germany aims to use artificial intelligence in industrial production and automation processes. With its plan called "National Strategy for Artificial Intelligence", it plans to invest 3 billion euros in artificial intelligence research and applications by 2025. The strategy also emphasizes the ethical and reliable use of artificial intelligence.

The United Kingdom

The UK has accepted artificial intelligence as a strategic priority and has taken important steps in this area. It aims to stand out in the global artificial intelligence race, especially by investing in artificial intelligence research and development activities. It has committed to investing in artificial intelligence research and education with the "Artificial Intelligence Sector Agreement". It has also established the "Artificial Intelligence Ethics and Innovation Centre" to provide guidance on the ethical use of artificial intelligence.

Switzerland

Switzerland is a world-renowned country for AI research, particularly in the fields of robotics and machine learning. The federal government is developing policies to promote the ethical and trustworthy use of AI.

Russia

Russia sees artificial intelligence as part of its national security and defense strategies. Having adopted the "National Strategy on Artificial Intelligence" in 2019, Russia aims to be among the leading countries in the field of artificial intelligence by 2030. President Vladimir Putin stated that artificial intelligence is important not only for Russia but for all of humanity.

India

India sees AI as a tool for economic development and social welfare. With its report titled "National Strategy for Artificial Intelligence," it plans to promote AI applications in areas such as agriculture, healthcare, education, and infrastructure.

Greece

Aware of the economic and social potential of AI, Greece is aligning itself with the EU's AI policies. The country is implementing various initiatives to support AI research.

Bulgaria

Bulgaria sees AI as part of the digital economy and is integrating into the EU's AI strategies. The country supports AI research and development activities.

Pakistan

Pakistan recognizes the potential of AI and aims to develop in this area. The government aims to develop skilled human resources in this field by investing in AI and data science education.

Iran

Iran considers artificial intelligence as a part of its science and technology policies. The country encourages artificial intelligence research and attaches importance to training talented human resources in this field.

Iraq

Iraq does not have a specific national strategy for artificial intelligence, but aims to develop in the areas of technology and digitalization.

Italy

Italy aims to use artificial intelligence in the industry and service sectors. With its "Artificial Intelligence Strategy", it supports artificial intelligence research and encourages its ethical use.